28
Topics in Organometallic Chemistry

Editorial Board:
M. Beller · J. M. Brown · P. H. Dixneuf · A. Fürstner
L. S. Hegedus · P. Hofmann · T. Ikariya · L. A. Oro ·
M. Reetz · Q.-L. Zhou

Topics in Organometallic Chemistry
Recently Published and Forthcoming Volumes

Molecular Organometallic Materials for Optics
Volume Editors: H. Le Bozec, V. Guerchais
Vol. 28, 2010

Conducting and Magnetic Organometallic Molecular Materials
Volume Editors: M. Fourmigué, L. Ouahab
Vol. 27, 2009

Metal Catalysts in Olefin Polymerization
Volume Editor: Z. Guan
Vol. 26, 2009

Bio-inspired Catalyst
Volume Editor: T. R. Ward
Vol. 25, 2009

Directed Metallation
Volume Editor: N. Chatani
Vol. 24, 2007

Regulated Systems for Multiphase Catalysis
Volume Editors: W. Leitner, M. Hölscher
Vol. 23, 2008

Organometallic Oxidation Catalysis
Volume Editors: F. Meyer, C. Limberg
Vol. 22, 2007

N-Heterocyclic Carbenes in Transition Metal Catalysis
Volume Editor: F. Glorius
Vol. 21, 2006

Dendrimer Catalysis
Volume Editor: L. H. Gade
Vol. 20, 2006

Metal Catalyzed Cascade Reactions
Volume Editor: T. J. J. Müller
Vol. 19, 2006

Catalytic Carbonylation Reactions
Volume Editor: M. Beller
Vol. 18, 2006

Bioorganometallic Chemistry
Volume Editor: G. Simonneaux
Vol. 17, 2006

Surface and Interfacial Organometallic Chemistry and Catalysis
Volume Editors: C. Copéret, B. Chaudret
Vol. 16, 2005

Chiral Diazaligands for Asymmetric Synthesis
Volume Editors: M. Lemaire, P. Mangeney
Vol. 15, 2005

Palladium in Organic Synthesis
Volume Editor: J. Tsuji
Vol. 14, 2005

Metal Carbenes in Organic Synthesis
Volume Editor: K. H. Dötz
Vol. 13, 2004

Theoretical Aspects of TransitionMetal Catalysis
Volume Editor: G. Frenking
Vol. 12, 2005

Ruthenium Catalysts and Fine Chemistry
Volume Editors: C. Bruneau, P. H. Dixneuf
Vol. 11, 2004

New Aspects of Zirconium Containing Organic Compounds
Volume Editor: I. Marek
Vol. 10, 2004

Precursor Chemistry of Advanced Materials
CVD, ALD and Nanoparticles
Volume Editor: R. Fischer
Vol. 9, 2005

Metallocenes in Stereoselective Synthesis
Volume Editor: T. Takahashi
Vol. 8, 2004

Molecular Organometallic Materials for Optics

Volume Editors: Hubert Le Bozec and Véronique Guerchais

With Contributions by

Zuqiang Bian · Marie P. Cifuentes · Santo Di Bella ·
Claudia Dragonetti · Nicholas C. Fletcher ·
Véronique Guerchais · Chunhui Huang ·
Mark G. Humphrey · M. Cristina Lagunas ·
Hubert Le Bozec · Zhiwei Liu · Lisa Murphy ·
Maddalena Pizzotti · Dominique Roberto ·
Marek Samoc · Francesca Tessore · Renato Ugo ·
J. A. Gareth Williams

Editors
Dr Hubert Le Bozec
Université de Rennes1
UMR CNRS 6226
Sciences Chimiques de Rennes
Campus de Beaulieu
35042 Rennes Cedex
France
hubert.le-bozec@univ-rennes1.fr

Dr Véronique Guerchais
Université de Rennes1
UMR CNRS 6226
Sciences Chimiques de Rennes
Campus de Beaulieu
35042 Rennes Cedex
France
Veronique.Guerchais@univ-rennes1.fr

ISSN 1436-6002   e-ISSN 1616-8534
ISBN 978-3-642-01865-7 e-ISBN 978-3-642-01866-4
DOI: 10.1007/978-3-642-01866-4
Springer Heidelberg Dordrecht London New York

Library of Congress Control Number: 2009927284

© Springer-Verlag Berlin Heidelberg 2010
This work is subject to copyright. All rights are reserved, whether the whole or part of the material is concerned, specifically the rights of translation, reprinting, reuse of illustrations, recitation, broadcasting, reproduction on microfilm or in any other way, and storage in data banks. Duplication of this publication or parts thereof is permitted only under the provisions of the German Copyright Law of September 9, 1965, in its current version, and permission for use must always be obtained from Springer. Violations are liable to prosecution under the German Copyright Law.

The use of general descriptive names, registered names, trademarks, etc. in this publication does not imply, even in the absence of a specific statement, that such names are exempt from the relevant protective laws and regulations and therefore free for general use.

Cover design: KünkelLopka GmbH; *volume cover:* SPi Publisher Services

Printed on acid-free paper

Springer is part of Springer Science+Business Media (www.springer.com)

Volume Editors

Dr. Hubert Le Bozec

Université de Rennes1
UMR CNRS 6226
Sciences Chimiques de Rennes
Campus de Beaulieu
35042 Rennes Cedex
France
hubert.le-bozec@univ-rennes1.fr

Dr. Véronique Guerchais

Université de Rennes1
UMR CNRS 6226
Sciences Chimiques de Rennes
Campus de Beaulieu
35042 Rennes Cedex
France
veronique.guerchais@univ-rennes1.fr

Editorial Board

Prof. Matthias Beller

Leibniz-Institut für Katalyse e.V.
an der Universität Rostock
Albert-Einstein-Str. 29a
18059 Rostock, Germany
matthias.beller@catalysis.de

Prof. John M. Brown

Chemistry Research Laboratory
Oxford University
Mansfield Rd.
Oxford OX1 3TA, UK
john.brown@chem.ox.ac.uk

Prof. Pierre H. Dixneuf

Campus de Beaulieu
Université de Rennes 1
Av. du Gl Leclerc
35042 Rennes Cedex, France
pierre.dixneuf@univ-rennes1.fr

Prof. Alois Fürstner

Max-Planck-Institut für Kohlenforschung
Kaiser-Wilhelm-Platz 1
45470 Mülheim an der Ruhr, Germany
fuerstner@mpi-muelheim.mpg.de

Prof. Louis S. Hegedus

Department of Chemistry
Colorado State University
Fort Collins, Colorado 80523-1872, USA
hegedus@lamar.colostate.edu

Prof. Peter Hofmann

Organisch-Chemisches Institut
Universität Heidelberg
Im Neuenheimer Feld 270
69120 Heidelberg, Germany
ph@uni-hd.de

Prof. Takao Ikariya

Department of Applied Chemistry
Graduate School of Science and
Engineering
Tokyo Institute of Technology
2-12-1 Ookayama, Meguro-ku,
Tokyo 152-8550, Japan
tikariya@apc.titech.ac.jp

Prof. H.C. Luis A. Oro

Instituto Universitario de Catálisis
Homogénea
Department of Inorganic Chemistry
I.C.M.A. - Faculty of Science
University of Zaragoza-CSIC
Zaragoza-50009, Spain
oro@unizar.es

Prof. Manfred Reetz

Max-Planck-Institut für Kohlenforschung
Kaiser-Wilhelm-Platz 1
45470 Mülheim an der Ruhr, Germany
reetz@mpi-muelheim.mpg.de

Prof. Qi-Lin Zhou

State Key Laboratory of Elemento-organic
Chemistry
Nankai University
Weijin Rd. 94, Tianjin 300071
PR CHINA
qlzhou@nankai.edu.cn

Topics in Organometallic Chemistry Also Available Electronically

Topics in Organometallic Chemistry is included in Springer's eBook package *Chemistry and Materials Science*. If a library does not opt for the whole package the book series may be bought on a subscription basis. Also, all back volumes are available electronically.

For all customers who have a standing order to the print version of *Topics in Organometallic Chemistry*, we offer the electronic version via SpringerLink free of charge.

If you do not have access, you can still view the table of contents of each volume and the abstract of each article by going to the SpringerLink homepage, clicking on "Chemistry and Materials Science," under Subject Collection, then "Book Series," under Content Type and finally by selecting *Topics in Organometallic Chemistry*.

You will find information about the

– Editorial Board
– Aims and Scope
– Instructions for Authors
– Sample Contribution

at springer.com using the search function by typing in *Topics in Organometallic Chemistry*.
Color figures are published in full color in the electronic version on SpringerLink.

Aims and Scope

The series *Topics in Organometallic Chemistry* presents critical overviews of research results in organometallic chemistry. As our understanding of organometallic structures, properties and mechanisms grows, new paths are opened for the design of organometallic compounds and reactions tailored to the needs of such diverse areas as organic synthesis, medical research, biology and materials science. Thus the scope of coverage includes a broad range of topics of pure and applied organometallic chemistry, where new breakthroughs are being made that are of significance to a larger scientific audience.

The individual volumes of *Topics in Organometallic Chemistry* are thematic. Review articles are generally invited by the volume editors.

In references *Topics in Organometallic Chemistry* is abbreviated Top Organomet Chem and is cited as a journal.

Preface

For many years, organometallic compounds have found widespread uses in organic synthesis, homogeneous catalysis and pharmaceutical chemistry. In the area of new molecular materials for photonic and optoelectronic applications, the impact of coordination and organometallic complexes of transition metals has also increased dramatically during the last two decades. Compared to organics, these compounds offer a large variety of molecular and supramolecular structures, a diversity of electronic and optical properties by virtue of the metal centre which can give rise to new functional and multifunctional materials. This book is intended to provide an overview of the development of organotransition metal complexes for optics. It describes the current state of art related to important areas such nonlinear optics, luminescence, sensing and photochromism.

The first two chapters are devoted to nonlinear optical activity of organometallic and coordination metal complexes which represent an emerging and growing class of NLO chromophores. The first chapter by S. Di Bella, C. Dragonetti, M. Pizzotti, D. Roberto, F. Tessore and R. Ugo presents an overview of the main classes of second-order NLO coordination and organometallic complexes with various ligands and is focused on NLO properties measured at the molecular level from solution studies, as well as on NLO properties of bulk materials. Significant recent advances in the third-order NLO properties of organometallic complexes are reviewed by M.G. Humphrey, M.P. Cifuentes and M. Samoc in the second chapter, with particular emphasis on spectral dependence studies and switching of nonlinearity.

There is a great potential in light-emitting diodes (OLEDs) for third-row transition metal complexes, as they represent an important class of triplet emitters. The emission colour is highly tuneable over a wide spectral range. In this context, it is of interest to provide an overview on the recent development of phosphorescent cyclometallated platinum and iridium complexes. In the third chapter, L. Murphy and J.A.G. Williams present the factors that need to be taken into account for a rational design of highly luminescent platinum complexes, and Z. Liu, Z. Bian and C.H. Huang focus on iridium complexes in the fourth chapter. Selected examples of the utility of these metal complexes in OLEDs and other applications are also included. Research in lighting systems is becoming one of the blooming fields nowadays and production of white light has been reported for a platinum complex, making it one of the most efficient WOLEDs reported hitherto.

Luminescent metal complexes have been shown to act as powerful reporters for molecular recognition and the contribution of M.C. Lagunas and N.C. Fletcher in the fifth chapter summarizes studies to date addressing the design and use of environment-responsive metal-based sensors. They highlight the modularity of complexes allowing the detection of various analytes, including protons, ions, oxygen, volatile organic compounds and small molecules. New research to elaborate molecular systems for addressing sensitivity, selectivity, aqueous compatibility and cellular uptake are expected to be developed in the near future.

Organic photochromic molecules are important for the design of photoresponsive functional materials, as switches and memories. Over the past 10 years, research efforts have been directed towards the incorporation of photoresponsive molecules into metal systems, in order either to modulate the photochromic properties, or to photoregulate the redox, optical and magnetic properties of the organometallic moieties. The sixth chapter by V. Guerchais and H. Le Bozec focuses on work reported within the last few years in the area of organometallic and coordination complexes containing photochromic ligands.

We are grateful to all the authors who have contributed to the volume in writing a chapter.

Rennes, France   Hubert Le Bozec
                                                                                                        Véronique Guerchais

Contents

Coordination and Organometallic Complexes as Second-Order Nonlinear Optical Molecular Materials 1
Santo Di Bella, Claudia Dragonetti, Maddalena Pizzotti, Dominique Roberto, Francesca Tessore, and Renato Ugo

NLO Molecules and Materials Based on Organometallics: Cubic NLO Properties 57
Mark G. Humphrey, Marie P. Cifuentes, and Marek Samoc

Luminescent Platinum Compounds: From Molecules to OLEDs 75
Lisa Murphy and J. A. Gareth Williams

Luminescent Iridium Complexes and Their Applications 113
Zhiwei Liu, Zuqiang Bian, and Chunhui Huang

Chromo- and Fluorogenic Organometallic Sensors 143
Nicholas C. Fletcher and M. Cristina Lagunas

Metal Complexes Featuring Photochromic Ligands 171
Véronique Guerchais and Hubert Le Bozec

Index 227

Coordination and Organometallic Complexes as Second-Order Nonlinear Optical Molecular Materials

Santo Di Bella, Claudia Dragonetti, Maddalena Pizzotti,
Dominique Roberto, Francesca Tessore, and Renato Ugo

Abstract Coordination and organometallic complexes with second-order nonlinear optical (NLO) properties have attracted increasing attention as potential molecular building block materials for optical communications, optical data processing and storage, or electrooptical devices. In particular, they can offer additional flexibility, when compared to organic chromophores, due to the presence of metal–ligand charge-transfer transitions, usually at relatively low-energy and of high intensity, tunable by virtue of the nature, oxidation state, and coordination sphere of the metal center. This chapter presents an overview of the main classes of second-order NLO coordination and organometallic complexes with various ligands such as substituted amines, pyridines, stilbazoles, chelating ligands (bipyridines, phenanthrolines, terpyridines, Schiff bases), alkynyl, vinylidene, and cyclometallated ligands, macrocyclic ligands (porphyrins and phthalocyanines), metallocene derivatives, and chromophores with two metal centers. The coverage, mainly from 2000 up to now, is focused on NLO properties measured at molecular level from solution studies, as well as on NLO properties of bulk materials.

Keywords Coordination and organometallic complexes, Second-order nonlinear optics

S. Di Bella (✉)
Dipartimento di Scienze Chimiche, Università di Catania, Viale A. Doria 8, 95125, Catania, Italy
e-mail: sdibella@unict.it

C. Dragonetti, M. Pizzotti, D. Roberto (✉), F. Tessore and R. Ugo
Dipartimento di Chimica Inorganica, Metallorganica e Analitica "Lamberto Malatesta" dell'Università degli Studi di Milano, UdR-INSTM di Milano and ISTM-CNR, Via Venezian 21, 20133 Milano, Italy
e-mail: dominique.roberto@unimi.it

Contents

1 Introduction 2
2 Principles of Second-Order Nonlinear Optics 3
　2.1　Basic Concepts and Methods 3
　2.2　Organic Molecular Materials 6
　2.3　Coordination and Organometallic Complexes 7
3 Coordination and Organometallic Complexes for Second-Order Nonlinear Optics 8
　3.1　Complexes with Amine, Pyridine, and Stilbazole Ligands 8
　3.2　Complexes with Chelating Ligands 13
　3.3　Complexes with Metallocene Ligands 23
　3.4　Complexes with Alkynyl and Vinylidene Ligands 26
　3.5　Cyclometallated Complexes 29
　3.6　Compounds with Macrocyclic Ligands 31
　3.7　Bimetallic Complexes 42
4 Conclusions and Perspectives 47
References 49

1 Introduction

Compounds with second-order nonlinear optical (NLO) properties are of great interest as molecular building block materials for optical communications, optical data processing and storage, or electrooptical devices [1–3]. Among them, organometallic and coordination metal complexes represent an emerging and growing class of second-order NLO chromophores that can offer additional flexibility, when compared to organic chromophores, due to the presence of metal–ligand charge-transfer (MLCT) transitions usually at relatively low energy and of high intensity, tunable by virtue of the nature, oxidation state, and coordination sphere of the metal center. Since the discovery of the second harmonic generation (SHG) for a ferrocenyl compound [4], increasing attention has been paid to organometallic and coordination complexes as potential second-order NLO chromophores. Extensive investigations have thus been carried out in this area. Early [5–9] and more recent [10–19] review articles on coordination and organometallic complexes with second-order NLO activity indicate the breadth of the active research in this field.

The goal of this overview is to report on the recent advances in order to analyze the main characteristics of second-order NLO organometallic and coordination complexes, and their potential as new NLO active molecular materials. After a brief introduction about the principles of nonlinear optics, this chapter illustrates the main classes of second-order NLO organometallic and coordination chromophores, producing examples of chromophores with monodentate nitrogen donor ligands (amines, pyridines, stilbazoles), chelating ligands (bipyridines, phenanthrolines, terpyridines, Schiff bases), alkynyl, vinylidene, and cyclometallated ligands, macrocyclic ligands (porphyrins and phthalocyanines), metallocene derivatives, and chromophores with two metal centers. The coverage, mainly from 2000 up to now, is not exhaustive, but allows the nonspecialist to get into this specific field and to understand its potentiality. Relevant older data of prototypical compounds are also reported. The focus is on both NLO properties measured at molecular level from solution studies and NLO properties of bulk materials.

2 Principles of Second-Order Nonlinear Optics

The principles of nonlinear optics, including the techniques to evaluate the second-order NLO properties, are briefly presented here. Major details can be found in excellent books [1–3] and various reviews [5–19].

2.1 Basic Concepts and Methods

Nonlinear optics deals with optical phenomena, caused by the interaction of applied electromagnetic fields to molecules or materials with emission of new electromagnetic fields which differ in frequency, phase, or other physical properties from the incident ones [1–3]. This kind of optical phenomena are related to the polarizability of a molecule or of a bulk material.

When a bulk material is subjected to an oscillating external electric field produced by an incident radiation, there is a polarization effect, expressed by

$$\vec{P} = \vec{P_0} + \vec{P}_{ind} = \vec{P_0} + \chi^{(1)}\vec{E}, \qquad (1)$$

where $\vec{P_0}$ is the intrinsic polarity, $\vec{P}_{ind}$ the induced polarization, and $\chi^{(1)}$ the electrical susceptibility or linear polarizability tensor. If the electric field strength $\vec{E}$ of the incident radiation is very high, as is the case with laser pulses, the perturbation is not linear and the induced polarization is better expressed by a power series according to

$$\vec{P} = \vec{P_0} + \chi^{(1)}\vec{E} + \chi^{(2)}\vec{E^2} + \cdots + \chi^{(n)}\vec{E^n}, \qquad (2)$$

where $\chi^{(2)}$, $\chi^{(3)}$, and $\chi^{(n)}$ tensors are, respectively, the second-, third-, and n-order electrical susceptibilities, controlling the nonlinear response of the material.

If, instead of a bulk material, the applied electromagnetic field is interacting with a molecule, the induced polarization is expressed by

$$\vec{P} = \mu_0 + \alpha\vec{E} + \beta\vec{E}^2 + \gamma\vec{E}^3 + \ldots, \qquad (3)$$

where μ_0 is the molecular ground state electric dipole moment, α the linear polarizability tensor, β and γ the non linear quadratic and cubic hyperpolarizability tensors, respectively, responsible for second- and third-order NLO effects. It is important to underline that both β and $\chi^{(2)}$ vanish in a centrosymmetric environment. Therefore, to have a second-order NLO effect, the acentricity requirement must be fulfilled. This is not true for γ and $\chi^{(3)}$.

The second-order NLO properties are of interest for a variety of NLO processes [1–3]. One of the most relevant is the SHG, originated by the mixing of three waves; two incident waves with frequency ω interact with the molecule or the bulk material with NLO properties, defined by a given value of the quadratic hyperpolarizability, β, or of the second-order electrical susceptibility, $\chi^{(2)}$, respectively, to produce a new electrical wave, named SH, of frequency 2ω. Another important second-order NLO process is the electrooptic Pockels effect which requires the presence of an external d.c. electric field, $E(0)$, in addition to the optical $\vec{E}(\omega)$ electrical field. This effect produces a change in the refractive index of a material proportional to the applied electric field, and can be exploited in devices such as optical switches and modulators [1–3].

To obtain molecular or bulk materials displaying significant second-order NLO effects, high values of β or of $\chi^{(2)}$, respectively, are required. In the case of molecules, in 1977 Oudar gave a theoretical interpretation of the electronic factors controlling β [20, 21]. The quadratic hyperpolarizability of a molecule is originated by the mobility of polarizable electrons under the effect of a strong electric field $\vec{E}$ associated with an incident radiation. It follows that it is dependent on electronic transitions which, being associated with a significant electronic mobility, are of high CT character. Oudar assumed that, when the second-order NLO response is dominated by one major CT process, β_{zzz} can be defined according to

$$\beta_{zzz} = \frac{3}{2h^2c^2} \frac{v_{eg}^2 r_{eg}^2 \Delta\mu_{eg}}{\left(v_{eg}^2 - v_L^2\right)\left(v_{eg}^2 - 4v_L^2\right)}, \tag{4}$$

where z is the axis of the direction of the CT, v_{eg} (cm^{-1}) the frequency of the CT transition, r_{eg} the transition dipole moment, $\Delta\mu_{eg}$ the difference between excited state μ_e and ground state μ_g molecular dipole moments, and v_L the frequency of the incident radiation. Equation (4) is the so-called "two level" model, a way to estimate the frequency dependent quadratic hyperpolarizability for specific types of second-order NLO chromophores, characterized by a single dominant CT transition. Extrapolation to zero frequency ($v_L = 0.0$ eV; $\lambda = \infty$) allows estimation, according to Eq. (5), of the static quadratic hyperpolarizability β_0, a useful figure of merit to evaluate the basic second-order NLO properties of a molecule:

$$\beta_0 = \beta_\lambda \left[1 - (2\lambda_{max}/\lambda)^2\right]\left[1 - (\lambda_{max}/\lambda)^2\right], \tag{5}$$

where β_λ is the quadratic hyperpolarizability value at λ incident wavelength and λ_{max} is the absorption wavelength of the controlling major CT. The molecular quadratic hyperpolarizability β can be expressed both in the cgs (cm^4 statvolt^{-1} = esu) or in the SI (C m^3 V^{-2}) unit systems (the conversion from the SI to the cgs system is given by the relation 10^{-50} C m^3 V^{-2} = 2.694 × 10^{-30} esu).

From the "two level" model it is possible to extrapolate the dipolar electronic requirements that a molecule must fulfill in order to show a significant second-order NLO response. It must be noncentrosymmetric, with CT transitions with large $\Delta\mu_{eg}$ and r_{eg} and at relatively low energy. This can be achieved, for instance, by separation of an electron-donor and an electron-acceptor group with a π-conjugated polarizable spacer, as occurs in classical 1D dipolar push–pull organic systems. Recently, multipolar systems, such as octupolar molecules, have been increasingly investigated, because it was shown that it is not only dipolar structures that may be the origin of significant SHG [1–3, 10–19].

From a theoretical point of view, various quantum mechanical methods allow the calculation of the molecular quadratic hyperpolarizability, β. Among them, the "sum over states" (SOS) approach also gives a useful way to define the electronic origin of the NLO response. Density functional theory (DFT) and time-dependent DFT (TD-DFT) or time-dependent HF (TD-HF) calculations [22, 23] are the most advanced theoretical methods. In particular, referring to the SOS method, it describes the tensor β_{ijk} in terms of all the electronic states interacting with the perturbing electric field, as an infinite expansion over a complete set of unperturbed excited states. Obviously, a simplification of this approach is the two-state model (Eq. 4) described above. These theoretical methods are a useful way of understanding hyperpolarizability–structure relationships, thus helping chemists to the design of new efficient molecular NLO chromophores.

Experimentally, mainly two techniques – the electric field induced second harmonic generation (EFISH) and hyper-Rayleigh scattering (HRS, also termed harmonic light scattering method) – are used in order to determine in solution the experimental value of the quadratic hyperpolarizability of molecular NLO chromophores.

The EFISH technique [24], suitable for dipolar neutral molecules, provides information on the molecular NLO properties through

$$\gamma_{\text{EFISH}} = (\mu\beta_\lambda/5kT) + \gamma(-2\omega;\omega,\omega,0), \tag{6}$$

where $\mu\beta_\lambda/5kT$ represents the dipolar orientational contribution, and γ (-2ω; ω, ω, 0), β_λ a third-order term at frequency ω of the incident wavelength, is the electronic contribution which is negligible for many molecules with a limited electronic polarizability. β_λ is the projection along the dipole moment axis of β_{VEC}, the vectorial component of the β_{ijk} tensor of the quadratic hyperpolarizability, working with an incident wavelength λ of a pulsed laser. To obtain the value of β_λ, it is thus necessary to know the value of the ground state dipole moment μ of the molecule. Moreover, in order to avoid overestimation of the quadratic hyperpolarizabilty due to resonance enhancements, it is necessary to choose an incident wavelength whose second harmonic is far from any electronic absorption of the molecule.

The HRS technique [25–27] involves the detection of the incoherently scattered second harmonic generated by the molecule in solution under irradiation with a laser of wavelength λ, leading to the mean value of the $\beta \times \beta$ tensor product. By analysis of the polarization dependence of the second harmonic signal, which can be evaluated selecting the polarization of the incident and scattered radiation, it is possible to obtain information about the single components of the quadratic hyperpolarizability tensor β. Unlike EFISH, HRS can also be used for ionic molecular species and for nondipolar molecules such as octupolar molecules. In this chapter, the quadratic hyperpolarizability measured with an incident wavelength λ by the EFISH and HRS techniques will be indicated as β_λ(EFISH) and β_λ(HRS), respectively.

Since the "two level" model (Eq. 4) applies well to NLO chromophores characterized by a major CT transition, the solvatochromic method may afford a way to evaluate the quadratic hyperpolarizability, but only the component along the major CT direction, β_λ(CT), by recording electronic absorption spectra of this absorption band in a series of solvents covering a wide range of dielectric constants and of refraction indexes [28]. This method, which does not require sophisticated instrumentation, can give a fair to good estimate of the quadratic hyperpolarizability, for instance in the case of 1D dipolar push–pull molecules, and at the same time it may allow the evaluation of the contribution of a given absorption band to the β value. This method may be more accurate in the case of fluorescent compounds, by combining the solvatochromic study of both absorption and emission spectra [28]. It should be emphasized that β_λ(EFISH) and β_λ(CT) values can be compared only when the dipole moment axis and the direction of the CT are roughly the same.

The static hyperpolarizability β_0 can also be evaluated by means of Stark (electroabsorption) spectroscopy, which affords the value of $\Delta\mu_{eg}$ by analyzing the effects of an applied electric field on the shapes of the major absorption bands of CT character [29–31]. The contribution of each transition to the quadratic hyperpolarizability value is thus obtained according to the "two level" model (Eq. 4). However, the estimated β_0 are generally approximate, especially when dynamic β data are resonantly enhanced, or when many excited states contribute to the NLO response [32].

In the case of bulk materials or films the second-order susceptibility values, $\chi^{(2)}$, can be obtained by means of the investigation of the SHG [1–3]. The Kurtz–Perry technique [33] is often used to compare the intensity of the SHG of a powder sample with that of a reference sample of known $\chi^{(2)}$, such as quartz or urea. Although this technique is limited (the magnitude of the response is also dependent on particle size), it is a simple and rapid method for screening a large number of powder materials.

2.2 Organic Molecular Materials

Although this chapter deals with molecular second-order NLO chromophores based on organometallic and coordination complexes, for the sake of clarity we report here a brief summary of the structural features of molecular organic second-order NLO chromophores, which have been extensively investigated [1–3]. In this way, we can give a description of the actual approach to the design of second-order NLO active chromophores.

As stated in the previous paragraph, the noncentrosymmetry is generally a prerequisite for second-order NLO activity of a molecule. However, in order to obtain efficient second-order molecular responses, intense, low-energy electronic transitions having CT character are required. With these concepts in mind, various synthetic strategies, through an appropriate molecular design, also recently based on sophisticated theoretical approaches, such as TD-DFT or TD-HF, have been developed in order to produce efficient second-order molecular NLO chromophores. Actually, for material chemists involved in the investigation of NLO properties, this is one of the most important topics of the latest two decades.

Two main families of organic molecular NLO chromophores can be identified: dipolar and octupolar species. The former, which are not centrosymmetric, follow a general scheme involving a polarizable molecular structure (e.g., a π-conjugated pathway) having an asymmetrical charge distribution (e.g., with donor and/or acceptor group substituents) to form a donor–π–conjugated bridge-acceptor (D–π–A) network. The prototypical example of a dipolar molecule is represented by p-nitroaniline. The second-order optical nonlinearity originates from the existence of D→A electronic CT transitions mediated by the π-conjugated-bridge, which in many cases are referred to the lowest-energy transition, so that the "two level" model applies quite well. To this category of molecular materials belong most conjugated organic species.

Octupolar molecules, instead, may be centrosymmetric but they imply the existence of twofold (D_2) or threefold (D_3) rotational axes. They are characterized by multidirectional CT excitations. The theoretical description of nonlinearity of such systems implies, even in the simplest case, a three-level approach. The prototypical example of an octupolar molecule is represented by 1,3,5-triamino-2,4,6-trinitrobenzene.

2.3 *Coordination and Organometallic Complexes*

Second-order NLO active coordination and organometallic complexes have progressively occupied in the last decade a relevant role in the panorama of molecular NLO chromophores because of the unique, structural and electronic characteristics associated with a metal center interacting with organic ligands [5–19]. Actually, compared to organic molecules, coordination and organometallic complexes can offer a larger variety of electronic structures, controlled by the metal electronic configuration, oxidation state, coordination sphere, etc. In particular, in the case of dipolar NLO chromophores, the metal center may act as the donor, or as the acceptor, or even as the bridge of a donor–acceptor network.

In fact, as coordination and organometallic complexes may possess intense, low-energy MLCT, ligand-to-metal CT (LMCT), or intraligand CT (ILCT) excitations, the metal can effectively act as the donor, the acceptor, or the polarizable bridge of a donor–acceptor network. Finally, metal ions are well suited to build molecular structures based on octupolar coordination of organic ligands with D_2 or D_3 symmetry.

In the following paragraphs is reported an overview, mainly limited to the last 8–10 years, on coordination and organometallic second-order NLO active chromophores, focusing the attention only on specific and relevant aspects associated with each class of NLO chromophores; therefore the overview is not exhaustive.

3 Coordination and Organometallic Complexes for Second-Order Nonlinear Optics

3.1 Complexes with Amine, Pyridine, and Stilbazole Ligands

3.1.1 Amines

Coe et al. investigated deeply the second-order NLO properties of various ruthenium amine complexes with pyridine ligands [18, 34–51]. In particular, they systematically investigated, in acetonitrile solution by the HRS technique working at 1.064 μm incident wavelength, Ru^{II} NLO chromophores such as *trans*-[Ru(NH$_3$)$_4$(L^D)(L^A)][PF$_6$]$_n$ ($n = 2$ or 3) where L^D (e.g., 4-(dimethylamino)pyridine) and L^A (e.g., 4-acetylpyridine or *N*-methyl-4,4′-bipyridinium, MeQ$^+$) are electron-rich and electron-deficient ligands, respectively [35]. In such chromophores, intense $Ru^{II} \rightarrow L^A$ MLCT transitions dominate the β(HRS) value. A rather high β_0(HRS) value was reported for the chromophores [Ru(NH$_3$)$_5$(MeQ$^+$)][PF$_6$]$_3$ (**1**, 123×10^{-30} esu) and *trans*-[Ru(NH$_3$)$_4$(4-Me$_2$N–C$_5$H$_4$N)(MeQ$^+$)][PF$_6$]$_3$ (130×10^{-30} esu). Complexes with the charged MeQ$^+$ ligand show larger β_0 values, when compared with their analogs with neutral pyridine L^A ligands, due to a more extended π-conjugation and to the presence of a ligand carrying a positive charge [35].

N-Arylation of the 4,4′-bipyridinium cation leads to an even higher β_0(HRS) value (**2**, 410×10^{-30} esu) [37], and, by placing a *trans*-ethylene bridge between the pyridine and pyridinium rings of L^A, β_0 increases up to 50% [41].

Besides, replacement of a neutral L^D ligand such as 4-Me$_2$N–C$_5$H$_4$N with a presumably N-coordinated thiocyanate anion (see **3**) increases the electron-donating strength of the Ru^{II} center and consequently the β_0(HRS) value to 513×10^{-30} esu [42]. The β_0 values obtained by the HRS technique for these Ru^{II} chromophores were confirmed by means of the method based on Stark spectroscopy [41, 42]. Also, calculations based on TD-DFT gave values of β_0 reasonably in agreement with the experimental β_0(HRS) values [46]. Interestingly, an [Ru(NH$_3$)$_5$NC$_5$H$_4$-]$^{2+}$ moiety is more effective than a 4-NMe$_2$–C$_6$H$_4$- moiety as a π-electron-donor, in terms of enhancing β_0(HRS) [47].

Within three series of pyridyl polyene chromophores **4**, the β_0 values obtained by the HRS technique or Stark spectroscopy (about 100–600 × 10^{-30} esu) maximize when $n = 2$. This is in contrast with the behavior of known organic chromophores based on an electron-donor–acceptor system linked by a polyene bridge, in which β_0(HRS) increases steadily with the increase of the π delocalization of the polyene bridge [48, 49]. TD-DFT calculations show that the HOMO level of these

Ru[II] complexes gains in π character as n increases; consequently, the lowest energy transition usually considered as purely MLCT in character has some opposite ILCT contribution which becomes significant when $n > 2$ and which increases with the conjugated pathlength, causing the value of β_0(HRS) to decrease [49]. Electrochemical studies, ^{1}H NMR, and an investigation based on Stark spectroscopy all confirm that the role of an extended conjugated π-system is more effective in the purely organic chromophores than in their Ru[II] analogs [50, 51].

Remarkably, the β_0(HRS) values of certain complexes (e.g., **1**) can be reversibly and very effectively (10- to 20-fold) attenuated by a Ru[III]/Ru[II] redox process involving chemical reagents such as H_2O_2 [39, 40, 43–45].

The redox-switching of the linear optical absorption of self-assembled monolayers and Langmuir–Schäfer films of [Ru(NH$_3$)$_5$(4,4'-bipyridinium)]$^{3+}$ complexes [52–54] and a redox-switching of the NLO response of Langmuir–Blodgett thin films based on **5** were recently reported. Oxidation to Ru[III] causes ca. 50% decrease of the intensity of the SHG, which is almost completely restored by reduction to Ru[II] [55].

3.1.2 Pyridines and Stilbazoles

The effect of coordination to a metal center on the second-order NLO response of pyridine and stilbazole ligands has been deeply studied mainly by Marks, Ratner et al., and by Ugo et al. [10–16]. The quadratic hyperpolarizability β, measured in solution by the EFISH technique, of *para*-substituted pyridines and stilbazoles, such as 4-R-C$_5$H$_4$N and 4,4'-*trans* or *trans,trans*-R-C$_6$H$_4$(CH=CH)$_n$–C$_5$H$_4$N (R = donor or acceptor substituent; $n = 1, 2$), increases upon coordination to various metal centers, the enhancement factor (EF) being modulated by the nature of the metal (electronic configuration, oxidation state, coordination sphere) which can act as an electron-acceptor or an electron-donor. For example, the quadratic hyperpolarizability, measured by the EFISH technique of 4,4'-*trans*-Me$_2$N–C$_6$H$_4$CH=CHC$_5$H$_4$N increases by a factor of 2 ($\beta_{1.06}$(EFISH) goes from 28.2×10^{-30} to 61×10^{-30} esu) upon coordination to a zerovalent "W(CO)$_5$" moiety which acts mainly as electron-acceptor [56]. Similarly, coordination of a pyridine or stilbazole ligand bearing an electron-withdrawing group to a low oxidation state organometallic fragment such as "M(CO)$_5$" (M=Cr, W), that may also act as electron-donor, could produce a relevant increase of the absolute value of the second-order NLO response. Thus, the $\beta_{1.91}$(EFISH) value of 4-COH–pyridine (-0.10×10^{-30} esu) increases its absolute value (-12×10^{-30} esu) upon coordination to a zerovalent "W(CO)$_5$" [56–58].

This ambivalent donor or acceptor role of a zerovalent metal has suggested two different mechanisms controlling the second-order NLO response of this kind of NLO organometallic chromophores [56] (Scheme 1).

When the R substituent is a strong electron-donating group, the increase of the value of the quadratic hyperpolarizability β is dominated by an intraligand CT (ILCT) transition, with the metal center, which behaves as an electron-acceptor,

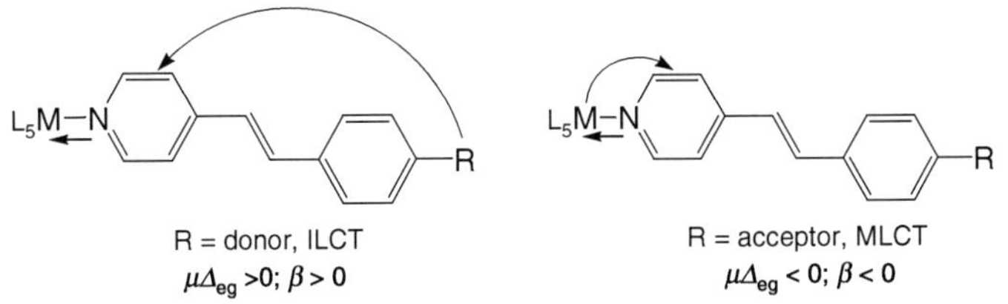

Scheme 1 Two mechanisms controlling the second order NLO response

producing a red-shift of this transition and therefore an increase of the value of β according to the "two level" model [20, 21]. In contrast, when R is a strong electron-accepting group, the quadratic hyperpolarizability β is dominated mainly by an MLCT transition. In this latter case the negative sign of β is due to a reduction of the dipole moment in the excited state of the MLCT transition ($\Delta\mu_{eg}$ <0), according to the "two level" model [20, 21].

An extended EFISH investigation of the second-order NLO response of cis-[M(CO)$_2$Cl(4-R-C$_5$H$_4$N)] (M = Rh, Ir) and fac-[Os(CO)$_3$Cl$_2$(4-R-C$_5$H$_4$N)] (R = electron-donor or -acceptor substituent) has confirmed such interpretation of the ambivalent role of the metal center. In fact, the ambivalent acceptor or donor role is controlled by the intrinsic softness of the metal center and by the presence in the pyridine ligand of an electron-donor or -acceptor group R [59]. For instance, the values of $\beta_{1.06}$(EFISH) of cis-[M(CO)$_2$Cl(4-R-C$_5$H$_4$N)] (M = Ir^I (5d^8), Rh^I (4d^8)) are dependent upon the nature of R: (1) β >0 with strong electron-donating groups (e.g., R = NMe$_2$ and M = Ir, $\beta_{1.06}$(EFISH) = 9 × 10^{-30} esu); (2) β is positive but very small with weak electron-donating groups (e.g., R = CMe$_3$ and M = Ir, $\beta_{1.06}$(EFISH) = ca. 0.1 × 10^{-30} esu); (3) β <0 with strong electron-withdrawing groups (e.g., R = CN and M = Ir, $\beta_{1.06}$(EFISH) = −9 × 10^{-30} esu). Independently from the strong electron-donating or withdrawing properties of the R group, a very strong enhancement of one or two orders of magnitude of the absolute value of $\beta_{1.06}$(EFISH) of the pyridine ligands occurs upon coordination [16].

Remarkably, it is possible to modulate the NLO response of a pyridine ligand by an increased acceptor or donor strength of the metal centers, which is controlled by their oxidation state and ancillary ligands [16]. Thus, coordination of 4-Me$_2$N–C$_5$H$_4$N to "cis-M(CO)$_2$Cl" (M = Rh^I, 4d^8 or Ir^I, 5d^8) or "fac-Os(CO)$_3$Cl$_2$" (Os^{II}, 5d^6) produces an increase of the $\beta_{1.06}$(EFISH) value about tenfold higher than coordination to the less accepting zerovalent "Cr(CO)$_5$" moiety (Cr^0, 3d^6). Besides, the quadratic hyperpolarizability $\beta_{1.91}$(EFISH) of [W(CO)$_5$(4-MeCO–C$_5$H$_4$N] (−9.3 × 10^{-30} esu) is larger than that of the complex of this substituted pyridine with the "Os^{II}(CO)$_3$Cl$_2$" moiety, due to the better donor properties of the "W(CO)$_5$" fragment [56, 57, 59]. Moreover, the value of $\beta_{1.34}$(EFISH) decreases upon substitution of the carbonyl ligands of "cis-M(CO)$_2$Cl" (M = Rh^I or Ir^I) with cyclooctadiene or with two cyclooctene ligands, according to a decrease of the acceptor properties of the metal center [59].

In push–pull 1D organic chromophores, an increase of the delocalized π-electron bridge between the donor and the acceptor groups leads to a significant increase of the second-order NLO response [60, 61]. In contrast, there is a buffering of the second-order NLO response upon coordination, because the effect due to coordination becomes less relevant by increasing the length of the π-delocalized bridge of the nitrogen donor ligand. For example, on going from 4-Me$_2$N–C$_5$H$_4$N to 4,4′-*trans*-Me$_2$N–C$_6$H$_4$CH=CHC$_5$H$_4$N the EF of $\beta_{1.34}$(EFISH) is of the order of 10^3 while on going from *cis*-[M(CO)$_2$Cl(4-Me$_2$N–C$_5$H$_4$N)] to *cis*-[M(CO)$_2$Cl(4,4′-*trans*-Me$_2$N–C$_6$H$_4$CH=CHC$_5$H$_4$N)] the EF is only ten times [59]. Coordination to "*cis*-Ir(CO)$_2$Cl" of 4,4′-*trans*-Me$_2$N–C$_6$H$_4$CH=CHC$_5$H$_4$N causes an EF of $\beta_{1.34}$(EFISH) of only 2.3, quite small when compared to the large EF produced when 4-Me$_2$N–C$_5$H$_4$N is coordinated to the same metal moiety [59].

Also, a zerovalent trimeric cluster core such as "Os$_3$(CO)$_{11}$" shows this ambivalent acceptor-donor effect on the second-order NLO response of *para*-substituted stilbazole ligands. However, strangely enough, its acceptor and donor strengths are comparable to those of "*fac*-Os(CO)$_3$Cl$_2$" and "*cis*-Ir(CO)$_2$Cl" [62]. While one expects a significant MLCT electron transfer from the "Os$_3$(CO)$_{11}$" core to the π* orbital system of a stilbazole ligand carrying in *para* position an electron-withdrawing substituent such as CF$_3$, in consideration of a significant polarizability of the *d* electron density of a trimeric cluster core with metals in the zero oxidation state, the significant σ acceptor properties of the zerovalent "Os$_3$(CO)$_{11}$" core when the stilbazole carries an electron-donor group such as NMe$_2$, comparable to that of an Os^{II} metal carbonyl center, is unexpected [62].

In contrast to the effect of coordination to a low oxidation state metal carbonyl moiety, the $\beta_{1.34}$(EFISH) value of 4,4′-*trans*-Me$_2$N–C$_6$H$_4$CH=CHC$_5$H$_4$N is almost unaffected by coordination to the relatively soft "*cis*-PtCl$_2$" moiety and, strangely enough, also to the less soft "Zn(CH$_3$CO$_2$)$_2$" Lewis acid moiety, in agreement with an irrelevant red-shift upon coordination of the ILCT transition of the stilbazole [63]. However, a more significant red-shift occurs upon coordination to harder "ZnY$_2$" (Y = Cl$^-$ [63], CF$_3$CO$_2^-$ [63], CF$_3$SO$_3^-$ [64, 65]) Lewis acid moieties. It appears that the β EF is function of the ancillary ligands Y which tune the acceptor properties of the Zn^{II} center (CH$_3$CO$_2^-$ <Cl$^-$ <CF$_3$CO$_2^-$ <CF$_3$SO$_3^-$), as confirmed by both EFISH measurements and solvatochromic investigations [63–65].

For instance, at concentrations higher than 1×10^{-4} M, the value of $\beta_{1.91}$(EFISH) of [Zn(CF$_3$SO$_3$)$_2$(4,4′-*trans*-Me$_2$N–C$_6$H$_4$(CH=CH)$_n$C$_5$H$_4$N)$_2$] (n = 1,2) is increased by a factor of about 4–6 times, when compared to that of the related complexes with the CH$_3$CO$_2^-$ ancillary ligand, in agreement with a higher Lewis acidity of the Zn^{II} center, as confirmed by the much higher red-shift of the ILCT transition upon coordination (for n = 1, $\Delta\lambda_{max}$ = 2 and 116 nm when Y = CH$_3$CO$_2^-$ and CF$_3$SO$_3^-$, respectively). Of interest is the unexpected evidence that, in CHCl$_3$ solution at concentrations lower than about 10^{-4} M, the $\beta_{1.91}$(EFISH) value of these triflate complexes increases abruptly by decreasing concentration, up to very large values, while this effect is not observed for the related acetate or trifluoroacetate complexes. Such behavior was attributed to an increased concentration of the cation [Zn(CF$_3$SO$_3$)(4,4′-*trans*-Me$_2$N–C$_6$H$_4$(CH=CH)$_n$C$_5$H$_4$N)$_2$]$^+$ (n = 1,2), characterized by a stronger second-order

NLO response due to the positive charge, produced by the facile solvolysis of the triflate ligand, as confirmed by electrical conductivity measurements which evidenced a sharp conductivity increase at concentrations below 10^{-4} M for the triflate complexes [64, 65]. A similar behavior of the quadratic hyperpolarizability in CHCl$_3$ solution by dilution was observed for Zn^{II} complexes with the same stilbazoles and with ancillary ligands such as the methansulfonate or *para*-toluensulfonate anions [66].

3.2 Complexes with Chelating Ligands

3.2.1 Bipyridines and Phenanthrolines

The second-order NLO response of various bipyridine and phenanthroline ligands increases upon coordination to a metal center, as in the case of pyridines and stilbazoles (see Sect. 3.1.2), the EF of the quadratic hyperpolarizability still depending on the electronic configuration of the metal, its oxidation state, and its sphere of ancillary ligands [10, 16].

An investigation was carried out on powders of Re, Pd, and Pt complexes with 2,2′-bipyridine, which exhibit modest second-order NLO activities as evidenced by the Kurtz technique [67]. A more extended investigation on the second-order NLO response of various Re^I, Zn^{II}, and Hg^{II} complexes with donor-substituted vinyl bipyridines (for examples, **6** and **7**) was done by the EFISH technique, working at 1.34 μm incident wavelength [10, 68, 69].

$\beta_{1.34}$(EFISH) increases with the strength of the electron-donor substituent group (NBu$_2$ is more efficient than Ooctyl) and with the Lewis acidity of the metal center (the relative increase of the acceptor strength, which parallels the increase of the red-shift of the ILCT transition upon coordination, follows the order: "Hg(OAc)$_2$" < "HgCl$_2$" < "Zn(OAc)$_2$" < "ZnCl$_2$" < "Re(CO)$_3$Br"), the best value being obtained for a complex with "ZnCl$_2$" (for **7** with M = Zn, $\beta_{1.34}$(EFISH) = 152 × 10^{-30} esu; $\mu\beta_0$ = 831 × 10^{-48} esu). The complexes with "Re(CO)$_3$Br" were less efficient than the corresponding Zn complexes due to the presence of two vectorially opposed CT transitions (MLCT and ILCT) [68, 69].

6 R = NBu$_2$, O = Ooctyl

7 M = Zn, Hg

In the related complexes **8**, for a given donor group, a slight increase of the value of $\mu\beta_{1.34}$(EFISH) occurs by replacing CH with N. However, the significant increase of the NLO response seems to be mainly controlled by resonant enhancement [70].

Interestingly, the $\mu\beta_{1.34}$(EFISH) value of **8a** [70] is twice that of the structurally related nonchelated complex [ZnCl$_2$(4,4′-*trans*-Me$_2$N–C$_6$H$_4$CH=CHC$_5$H$_4$N)$_2$] [63], due to the rather planar arrangement of the chelated ligand upon coordination and consequently to a shift of the ILCT transition at lower energy [16].

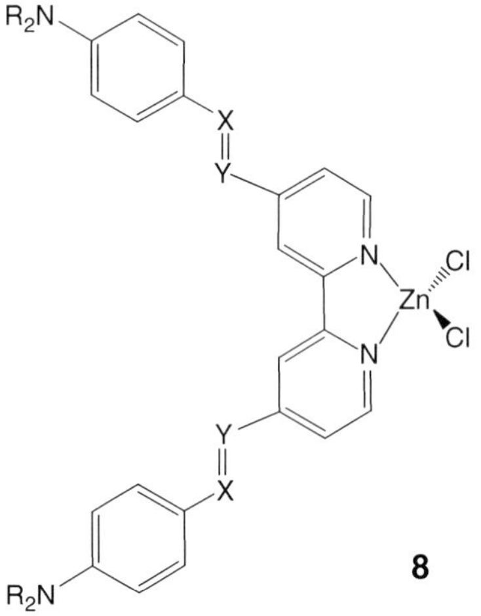

a X = Y = CH, R = Bu
b X = Y = CH, R = Et
c X = Y = N, R = Bu
d X = N, Y = CH, R = Et

8

Very recently, the efficient photoswitching of the second-order NLO responses of some dipolar photochromic Zn^{II} complexes (**9**) was reported [71]. The NLO activity for the open forms is very small, as expected for the absence of π-conjugation between the two thiophene rings. Upon conversion to the closed form in the photo-stationary state, the NLO activity increases dramatically ($\mu\beta_0$(EFISH) from 75–160 × 10^{-48} to 1,020–1,800 × 10^{-48} esu). This substantial enhancement reflects a more efficient delocalization of the π-electron system in the closed forms favoring an efficient on/off switching of the NLO responses [71].

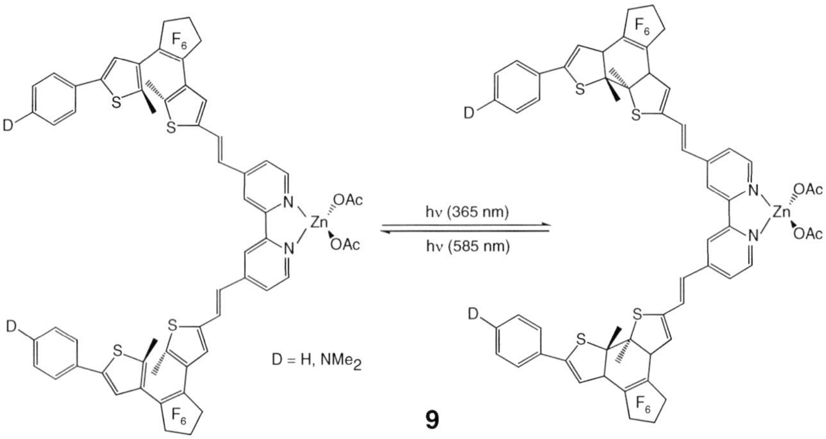

hν (365 nm)
hν (585 nm)

D = H, NMe$_2$

9

Two-dimensional second-order NLO, positively charged, chromophores incorporating the "$Ru(NH_3)_4{}^{2+}$" moiety linked to two structurally related electron-accepting *N*-methyl/aryl-pyridinium systems (either two monodentate pyridine ligands or one bipyridine chelated ligand) were studied (**10**) [72]. The $\beta_{0.80}$(HRS) for the chelated complexes (32–48 $\times$ 10^{-30} esu) are smaller than those of the nonchelated counterparts (135–160 $\times$ 10^{-30} esu). The possibilities for redox-switching of the NLO properties, by exploiting reversible $Ru^{III/II}$ oxidation processes and ligand-based reductions, were investigated [72].

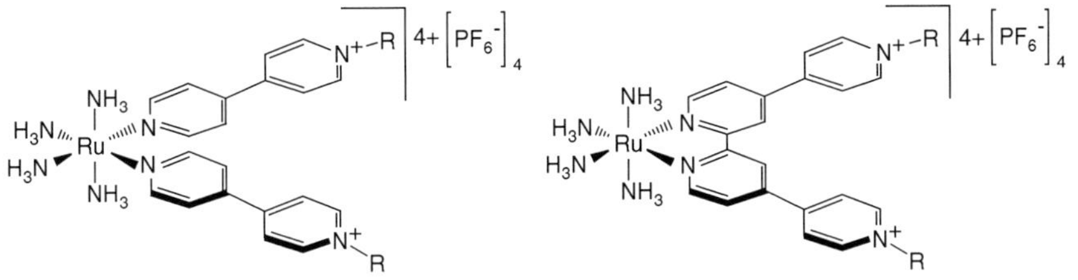

R = Me, Ph, 4-Acetyphenyl

R = Me, Ph, 4-Acetyphenyl

10

The quadratic hyperpolarizability of 5-R-1,10-phenanthroline (R = donor group such as OMe, NMe_2, *trans*-CH=CHC_6H_4-4′-NMe_2 and *trans,trans*-$(CH{=}CH)_2C_6H_4$-4′-NMe_2) increases upon coordination to a Zn^{II} center but not to the softer Cd^{II} center [63]. The EF of $\beta_{1.34}$(EFISH) upon coordination to "$Zn(CH_3CO_2)_2$" is, as expected, higher for the donor group NMe_2 (EF = 4.6) than for the donor group OMe (EF = 3.2). Such enhancement becomes less and less relevant by increasing the length of the π-delocalized bridge between the donor group NMe_2 and the phenanthroline chelated ligand, as already observed in the case of 4-R-pyridines (Sect. 3.1.2) [63]. Notably, the second-order NLO chromophore **11** (R = *trans,trans*-$(CH{=}CH)_2C_6H_4$-4′-NMe_2) is characterized by a good value of $\beta_{1.34}$(EFISHG) (112 $\times$ 10^{-30} esu) [63].

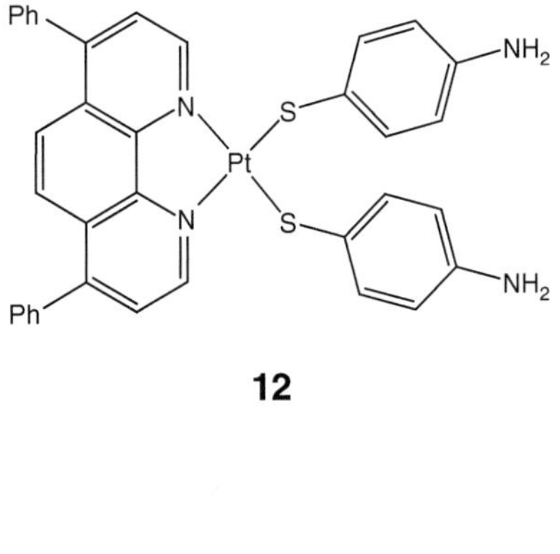

12

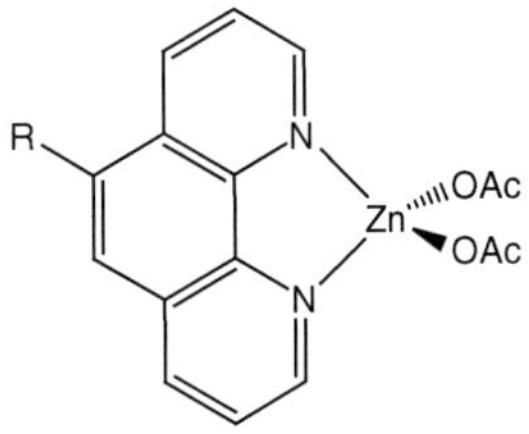

11

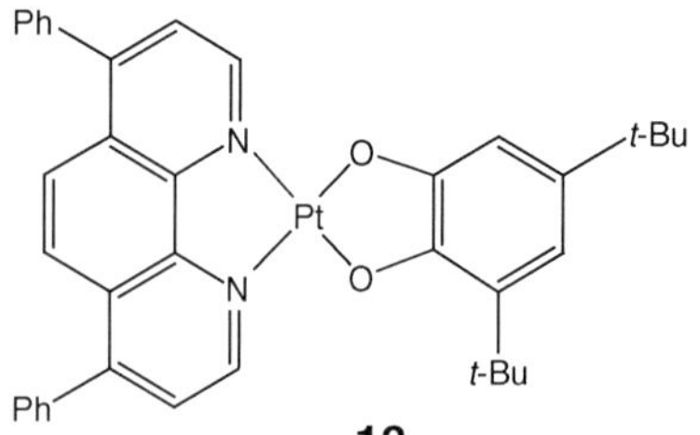

13

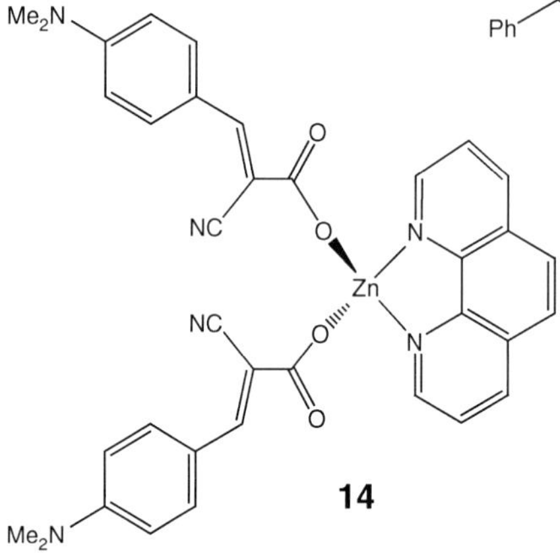

14

The $\beta_{1.91}$(EFISH) value of a large series of square planar M(substituted 1,10-phenanthroline)(dithiolate) complexes (M = Pt^{II}, Pd^{II}, Ni^{II}) was measured [73]. In these second-order NLO chromophores, the HOMOs are a mixture of metal and thiolate orbitals whilst the LUMOs are phenanthroline π* orbitals. Therefore, the presence of an electron-donor group on the dithiolate ligand, acting on the HOMOs level, increases the value of the quadratic hyperpolarizability by acting on the energy of the CT involving the HOMO–LUMO transition. The best NLO response is obtained when M = Pt^{II} (**12**, $\beta_{1.91}$(EFISH) = −39 × 10^{-30} esu; $\mu\beta_0$ = −260 × 10^{-48} esu) since this latter Pt^{II} complex shows an increased oscillator strength of the HOMO–LUMO transition when compared to that of the related Pd^{II} or Ni^{II} NLO complexes. The structurally related complex **13**, which shows a $\mu\beta_0$ value between −523 and −669 × 10^{-48} esu, dependent on the nature of the solvent, was also investigated [74].

The quadratic hyperpolarizability of various one-dimensional push–pull carboxylate ligands increases upon coordination to a "(1,10-phenanthroline)Zn^{II}" center. For example, **14** has a β_0(HRS) value of 39 × 10^{-30} esu, much higher than that of the related carboxylate ligand (6.6 × 10^{-30} esu) [75].

Metal ions are excellent templates to build D_{2d} and D_3 octupolar second-order NLO chromophores; therefore their associations with functionalized bipyridyl ligands give rise to octahedral and tetrahedral metal complexes with large quadratic hyperpolarizabilities [76]. The adequate functionalization of these octupoles allows their incorporation into macromolecular architectures such as polymers and metallodendrimers [76]. Various D_3 octupolar tris chelated Ru^{II} complexes are characterized by a significant second-order NLO response as evidenced by HRS [18].

For example, the β_0(HRS) value of **15**, when M = Ru^{II} and R = NBu_2, is 380 × 10^{-30} esu [77, 78]. Similar values are obtained upon substitution of PF_6^- with other anions such as $TRISPHAT^-$ (tris(tetrachlorobenzendiolato)phosphate anion) [79]. Polarized HRS investigations and Stark spectroscopic measurements support a second-order NLO response of **15** controlled by multiple degenerate dipolar CT transitions, rather than by an octupolar transition. In accordance the transitions dominant on the NLO response are ILCT excitations red-shifted by coordination to Ru^{II} and MLCT transitions with a CT process opposite as direction [78, 80].

Holding fixed the bipyridine carrying R = NBu_2, but changing the metal, substitution of Ru^{II} with Fe^{II} causes a decrease of β_0(HRS) due to both a blue-shift of the ILCT and a red-shift of the MLCT [78, 81]. The analogs Zn^{II} or Hg^{II} complexes, which have only the ILCT transition, show β_0(HRS) values of 380 and 256 × 10^{-30} esu, respectively, in accordance with the different Lewis acidity of the metal center [78]. By introducing the less donor R = Ooctyl group in the Ru^{II} NLO chromophore, the β_0(HLS) value decreases due to a ILCT absorption band at higher energy [78]. The NLO chromophore **15**, with M = Zn^{II} and R = *trans*-$CH=CHC_6H_4NBu_2$, exhibits the highest value of the quadratic hyperpolarizability reported for an octupolar NLO chromophore (β_0(HRS) = 657 × 10^{-30} esu; $\beta_{1.91}$(HRS) = 870 × 10^{-30} esu) [82].

Tetrahedral D_{2d} octupolar metal NLO chromophores (where the metal center M is Cu^I, Ag^I, Zn^{II}) with the same functionalized bipyridyl ligands were studied. Their quadratic hyperpolarizabilities, measured by HRS at 1.907 μm incident wavelength, are much lower when compared to those of the related octahedral D_3 NLO chromophores (β_0(HRS) = 70–157 × 10^{-30} esu and 200–657 × 10^{-30} esu for D_{2d} and D_3 complexes, respectively) [82].

15

M = Ru^{II}, Zn^{II}, Hg^{II}, Fe^{II}
R = NBu_2, Ooctyl, *trans*-$CH=CHC_6H_4NBu_2$

Le Bozec and coworkers have incorporated the Ru^{II} NLO chromophore **15**, adequately functionalized with an appropriate R = amino group, into macromolecular systems of interest as building blocks for the preparation of second-order NLO active bulk materials [83–86]. A thermally stable polyimide derivative of **15** shows, in CH$_2$Cl$_2$ solution, a $\beta_{1.91}$(HRS) value of 1,300 × 10^{-30} esu, larger than that of the monomeric counterpart [83], whereas a dendrimeric species **16** with seven units shows a $\beta_{1.91}$(HRS) value of 1,900 × 10^{-30} esu (in CH$_2$Cl$_2$) [84, 85]. The quasi-optimized octupolar ordering of the dendrimeric structure is responsible for the greater second-order NLO response compared to that of a linear arrangement of 14 basic units of the polyimide derivative [84, 85].

A star-shaped arrangement based on an octupolar Zn^{II} second-order NLO chromophore containing three photoisomerizable ligands (in two geometries) such as 4,4'-bis-(styryl)-2,2'-bipyridine functionalized with a dialkylamino-azobenzene shows a $\beta_{1.91}$(HRS) value of 863 × 10^{-30} esu in CH$_2$Cl$_2$ solution [87]. By using both photophysical and second-order NLO properties of this chromophore, all-optical poling, an interference process between one- and two-photon excitations that locally induces macroscopic second-order effects in polymeric films, was investigated. Grafting these chromophores onto the polymer network improves the stability of the macroscopic photoinduced nonlinearity [88].

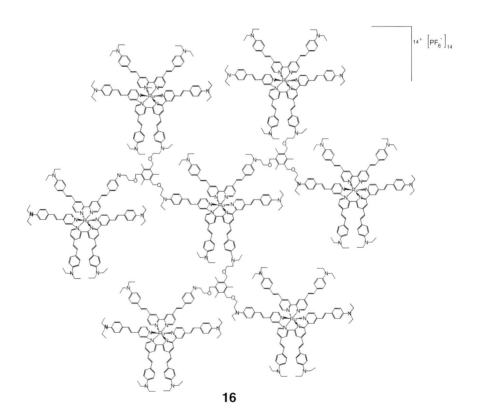

16

Coe et al. investigated some related tris-chelate complexes of Ru^{II} and Fe^{II}, with bipyridines substituted with electron-withdrawing pyridinium groups (**17**) [89]. In these particular octupolar chromophores the second-order NLO response is dominated by two MLCT transitions; HRS measurements working with a 0.800 µm incident wavelength have produced larger quadratic hyperpolarizabilities for the Ru^{II} chromophores (170–290 × 10^{-30} esu) than for the Fe^{II} analogs (78–110 × 10^{-30} esu). However, this is probably due to the presence of a resonant enhancement in the case of the Ru^{II} chromophores. A Stark spectroscopic investigation and computations based on TD-DFT theory and the finite field method have shown in fact that the second-order NLO response of these chromophores is controlled by two substantial components producing slightly larger β_0 values for the Fe^{II} chromophore (86–151 × 10^{-30} esu) than for the Ru^{II} chromophore (68–114 × 10^{-30} esu) [89]. Since MLCT transitions determine the second-order NLO response of these chromophores, NLO redox switching based on these Fe^{II}/Fe^{III} or Ru^{II}/Ru^{III} species is possible.

M = Fe, Ru
R = Me, Ph, 4-Acetyphenyl

17

3.2.2 Terpyridines

The second-order NLO properties of metal complexes with terpyridine ligands were recently studied for the first time [90, 91]. Coordination of a terpyridine such as 4′-(1-C$_6$H$_4$-p-NBu$_2$)-2, 2′: 6′, 2″-terpyridine to Zn^{II}, Ru^{III}, and Ir^{III} metal centers (**18**) induces a significant enhancement of its second-order NLO response measured by the EFISH technique working with an incident wavelength of 1.34 µm [90, 92]. In all these chromophores the ILCT transition of the terpyridine at about 416–465 nm is red-shifted by coordination, due to an increased π delocalization of the ligand upon chelation. The enhanced value upon coordination of the terpyridine to the Lewis center "ZnY$_2$" ($\beta_{1.34}$(EFISH) = 67 and 88 × 10^{-30} esu for Y = Cl$^-$, CF$_3$CO$_2^-$, respectively) remains positive, as expected for an ILCT transition dominating the second-order NLO response [16], the EF being higher for the more electron-withdrawing ancillary ligand CF$_3$CO$_2^-$. However, coordination of these terpyridine metal centers with different d^n configurations such as

Ir[III] ($5d^6$ in "Ir(4-EtPhCO$_2$)$_3$") or Ru[III] ($4d^5$ in "Ru(CF$_3$CO$_2$)$_3$"), produces an NLO response no longer influenced just by the ILCT transition, but also by MLCT and/or LMCT transitions, in such a way that these latter transitions can even change the sign of the quadratic hyperpolarizability. This relevant role of MLCT and LMCT transitions was confirmed by solvatochromic [92] and TD-DFT investigations [93]. As expected, the relevance of the MLCT transition increases when the terpyridine is carrying an electron-withdrawing substituent ($\beta_{1.34}$(EFISH) from -70 to -230×10^{-30} esu for R = NBu$_2$ and NO$_2$, respectively, when the terpyridine is coordinated to "Ir(4-EtPhCO$_2$)$_3$") [92].

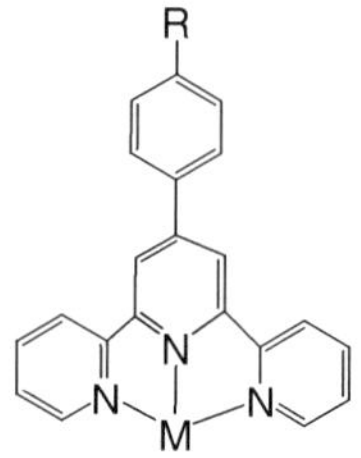

18

R = NBu$_2$, NO$_2$, (E)-CH=CH-C$_6$H$_4$-p-NBu$_2$,
(E),(E)-(CH=CH)$_2$-C$_6$H$_4$-p-NMe$_2$,

M = ZnCl$_2$, Zn(CF$_3$CO$_2$)$_2$, RuCl$_3$,
Ru(CF$_3$CO$_2$)$_3$, IrCl$_3$, Ir(4-EtPhCO$_2$)$_3$

Langmuir–Blodgett films of Zn[II] and Ir[III] complexes of 4'-(1-C$_6$H$_4$-p-NMe (C$_{16}$H$_{33}$))-2, 2': 6', 2''-terpyridine have been investigated, measuring their SHG at 1.064 μm incident wavelength. Rather low $\chi^{(2)}$ values were unexpectedly obtained, probably due to a scarce noncentrosymmetric ordering of these rather bulk chromophores [94].

Recently, the second-order NLO properties of new lanthanide complexes of the type [Ln(NO$_3$)$_3$–L] (Ln = La, Gd, Dy, Yb, Y; **19**), where L is a rather rigid terpyridine-like ligand, have been determined by HRS, working with a nonresonant incident wavelength of 1.907 μm. The value of the quadratic hyperpolarizability $\beta_{1.91}$(HRS) increases by increasing the number of f-electrons, from 186 to 288 $\times 10^{-30}$ esu [95]. The dependence of the second-order NLO response on the nature of the lanthanide metal center suggests that f-electrons may contribute to the second-order NLO response [95].

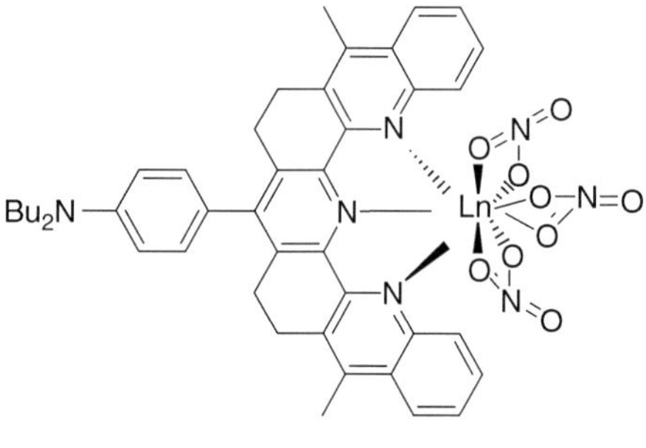

19

3.2.3 Schiff-Bases

Schiff-bases, arising from condensation of substituted salicylaldehydes with various bridging diamines, represent suitable templates to generate noncentrosymmetric molecular architectures. Various bis(salicylaldiminato)M^{II} (M = Fe, Co, Ni, Cu, Zn) complexes have been investigated as second-order NLO molecular chromophores [96, 97], exploring various aspects of their second-order nonlinear optics. In these NLO chromophores the metal ion templates noncentrosymmetric structures and acts as donor counterpart of a donor (D)–acceptor (A) system involving MLCT transitions. Starting from the more simple unsubstituted species (**20**), for the first time the role of the metal-d configuration was demonstrated in controlling the second-order NLO response, when MLCT transitions are dominating such a response. In particular, on passing from the closed-shell Ni^{II} (d^8) ($\beta_{1.34}$(EFISH) = -20×10^{-30} esu) to the open-shell Cu^{II} (d^9) ($\beta_{1.34}$(EFISH) = -50×10^{-30} esu) and Co^{II} (d^7) ($\beta_{1.34}$(EFISH) = -170×10^{-30} esu) homologues, the substantial increase of the absolute values of the quadratic hyperpolarizability is clearly related to the increased accessibility of lower-lying CT states [98, 99]. However, in the presence of strong donor/acceptor substituents on the Schiff-base structure, the metal ion mostly acts as a bridge, and its role in controlling the second-order NLO response becomes less defined. Tunable values of the quadratic hyperpolarizability, ranging from negative to positive values [97], were obtained for these molecular NLO chromophores, depending upon the strength of the donor/acceptor substituents. The symmetric donor–acceptor substitution in dipolar planar Schiff-base complexes also allowed investigation of the in-plane two-dimensional (2D) NLO properties (**21**), of interest for the development of polarization-independent materials.

M = Co, Ni, Cu
20

perpendicular
β_{zxx}, β_{xzx}
21

parallel
β_{zzz}

$\mu(Z)_{ge}$
$\mu(X)_{ge}$

Dipolar 1D donor/acceptor systems are generally characterized by a prevalent single hyperpolarizability tensor, namely β_{zzz}, parallel to the dipolar z axis of the molecule. However, an appropriate donor/acceptor substitution pattern on the bis(salicylaldiminato) framework, resulting in a C_{2v} molecular symmetry, leads to large off-diagonal β_{ijk} tensors components (e.g., **21**, D = NEt$_2$; A = Cl, $\beta_{zzz(1.34)}$(HRS) = 43×10^{-30} esu; $\beta_{zxx(1.34)}$(HRS) = 20×10^{-30} esu) [100], due to the existence of CT transitions (**21**), perpendicular to the two axes, $\mu(x)_{ge}$ [100, 101].

A further variety of noncentrosymmetric Schiff-base structures can be envisaged, either by an unsymmetrical donor/acceptor substitution on the bis(salicylidene) ligand of a series of Cu^{II} complexes (**22**) [102], or by an unsymmetrical [N$_2$O$_2$] tetradentate coordination in M^{II} (M = Ni, Cu, Zn, VO) complexes (**23**) [103], or finally in a series of Ni^{II} complexes with ligands derived by a monocondensation of the bridging diamine [104]. Appreciably optical nonlinearity has been achieved (e.g., **23**, M = Zn; R = N$_2$Ph, $\beta_{1.91}$(EFISH) = -280×10^{-30} esu) [103]. Therefore a large diversity of structures can produce second-order nonlinearity in this class of NLO chromophores. Further examples of second-order NLO chromophores based on Schiff bases are represented by a series of octahedral M^{II} (M = Fe, Co, Ni, Zn) metal complexes of *N*-2'-pyridylmethylene-4-aminopoly (phenyl) ligands, in which the molecular quadratic hyperpolarizability is strongly influenced by the metal electronic configuration and parallels the number of unpaired electrons. The largest values of the quadratic hyperpolarizability are those of the Mn^{II} complexes with a $3d^5$ electronic configuration [105]. Moreover, the switching of the second-order NLO response can be envisaged in the case of spin-crossover of Fe^{II} complexes. Bimetallic NLO chromophores involving dicopper(II) [106] or Cu^{II}-Gd^{III} [107] complexes have been investigated with the aim of finding possible interplay between magnetism and second-order NLO response. These "multifunctional" molecular chromophores, even possessing a rather modest optical nonlinearity, represent interesting models for further investigations of this field of research.

The use of chiral diamines in Schiff bases complexes allowed exploration of the powder SHG efficiency of crystalline materials based on this class of NLO chromophores, otherwise inactive because of the almost always crystal centrosymmetry. Relatively high powder SHG efficiency (up to 13 times that of urea) has been achieved in the case of the (1*R*,2*R*)-(+)-1,2-diphenylethylenediamine Ni^{II} derivative [108]. Analogous Zn^{II} complexes using the chiral (*R*)-(+)-1-phenylethylamine have given appreciable powder SHG efficiencies [109]. This strategy has been extended using a series of chiral amino alcohols [110] and amino acids [111] to obtain noncentrosymmetric crystals based on Sn^{IV} derivatives, with an attempt to correlate their SHG efficiencies with the molecular chirality.

In spite of the above studies on molecular NLO chromophores, investigations as NLO materials based on these chromophores remain almost unexplored. Monolayers of Ni^{II} complexes on glassy [112] or Si(100) [113] substrates have been obtained, but their second-order NLO activity has not yet been investigated.

3.3 Complexes with Metallocene Ligands

Since the first report in 1987 by Green et al. [4], metallocene derivatives represent one of the most widely investigated classes of second-order NLO metal-based chromophores [114–116]. Starting from the prototypical stilbene (*trans*)-1-ferrocenyl-2-(4-nitrophenyl)ethylene derivative (**24**, $\beta_{1.91}$(EFISH) = 31 × 10^{-30} esu) [117] and related phenylethenyl oligomers [118], a very large variety of ferrocenyl species have been investigated. They range from various polyenes having a terminal acceptor group [119], such as *N*-alkylpyridinium salts (e.g., **25**, $\beta_{1.06}$(HRS) = 458 × 10^{-30} esu) [120], fullerene [121], sesquifulvalene (e.g., **26**, $\beta_{1.06}$(HRS) = 1,539 × 10^{-30} esu; β_0(HRS) = 821 × 10^{-30} esu) [122], fluorene (e.g., **27**, R = CO$_2$Me, $\mu\beta_{1.54}$(EFISH) = 5,000 ± 1,500 × 10^{-48} esu) [123], indanone [124], thiazole [125], dicyanomethylene (e.g., **28**, $\mu\beta_{1.00}$(EFISH) = 1,120 × 10^{-48} esu), [126] and related derivatives [127–129], including some (dicyanomethylene)indane species (e.g., **29**, $\mu\beta_{1.91}$(EFISH) = 5,200 (8,720) × 10^{-48} esu) [130], and various bimetallic (see Sect. 3.7) [131–133] and trimetallic [134] compounds, connected to the ferrocenyl unit through a conjugated π-linker. The values of the quadratic hyperpolarizability of these species parallel the strength of the acceptor group and, generally, increase with the length of conjugated π-network, as usually observed for related organic chromophores [60, 61]. In all cases, the metallocene unit represents the donor group of the donor-acceptor system connected by a π linker. Actually, the ferrocene unit possesses ionization energy and redox potential features comparable to those of the best organic donors. However, the values of the quadratic hyperpolarizability of the NLO chromophores based on ferrocene as donor group indicate a donor capability of the ferrocenyl group comparable to that of the poorly donor organic methoxyphenyl group. These relatively poor donor properties are due to a weak electronic coupling between the metal-*d* orbitals of the metallocenyl donor group and the π-network of the linker connecting the donor-acceptor push-pull systems. Thus, the very large hyperpolarizability values observed for some of the NLO chromophores reported above are in some cases due to resonant enhancement effects. On the other hand, even extrapolated β_0 values [117–134] cannot be considered reliable [32], since it has been assessed that, for ferrocenyl derivatives, many electronic states contribute to optical nonlinearity [115, 116].

Some related ruthenocene species have also been investigated [117], almost always showing lower quadratic hyperpolarizabilities, in accordance with the higher ionization energy of the ruthenocene vs ferrocene moiety [115].

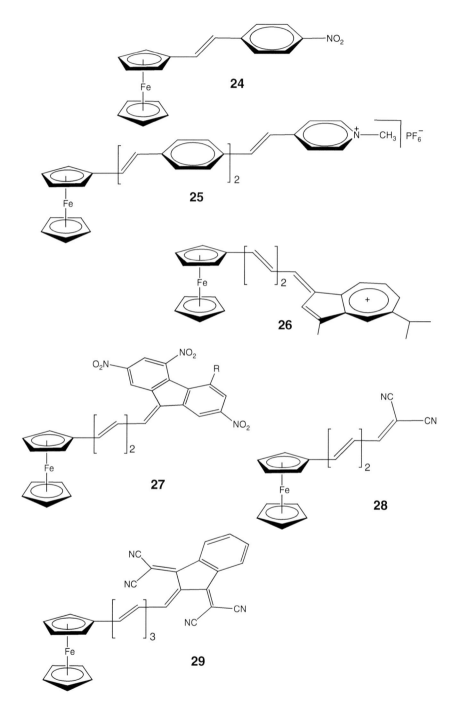

The interest in this class of NLO chromophores is also due to their capability to crystallize in noncentrosymmetric structures characterized by a high SHG, because the acceptor substituents are usually bulky species [135] or pyridinium salts [4, 136] which favor noncentrosymmetric packing. Also relevant is the possibility of engineering crystalline structures through chirality of the ferrocenyl group (e.g., **30**, powder efficiency 100 times that of urea) [137], or by an appropriate substituent on the second cyclopentadienyl ring (e.g., **31**, powder efficiency 140 times that of urea) [138]. Actually, the crystalline species based on ferrocene NLO chromophores are among the most SHG efficient organometallic materials, e.g., the pyridinium derivative **32** possesses powder efficiency 220 times that of urea [136].

The large variety of molecular NLO chromophores based on metallocenes rends them suitable for the investigation of various bulk second-order NLO materials. Thus, some studies devoted to electrically poled polymers including a metallocene NLO molecular chromophore as guest [126, 139], or to self-assembled nanostructures of these metallocene NLO chromophores [140, 141], appeared quite recently in the literature. Moreover, the facile and reversible accessibility to the ferrocenium oxidized species [142], allowed the redox switching of the SHG signal in self-assembled monolayers [140] or the molecular multifunctional (linear optical, NLO, and magnetic properties) redox switching [143] to be obtained.

3.4 Complexes with Alkynyl and Vinylidene Ligands

(Cyclopentadienyl) (alkylphosphine) metal σ-acetylides of group 8, group 10, and group 11 metals represent a widely investigated class of second-order NLO chromophores, mainly developed by Humphrey et al., whose relevant results are summarized in recent review articles [11, 19, 144].

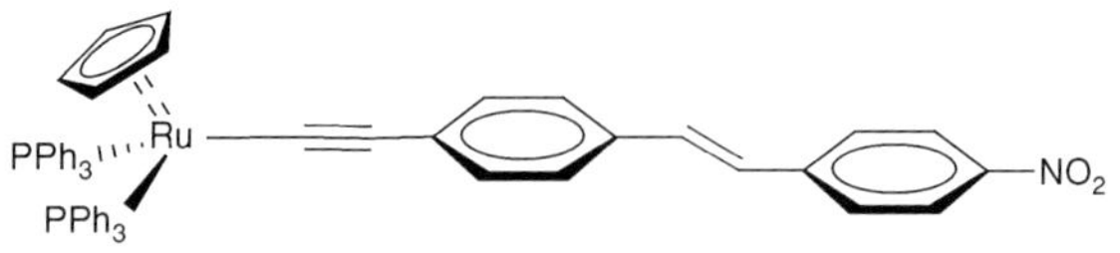

33

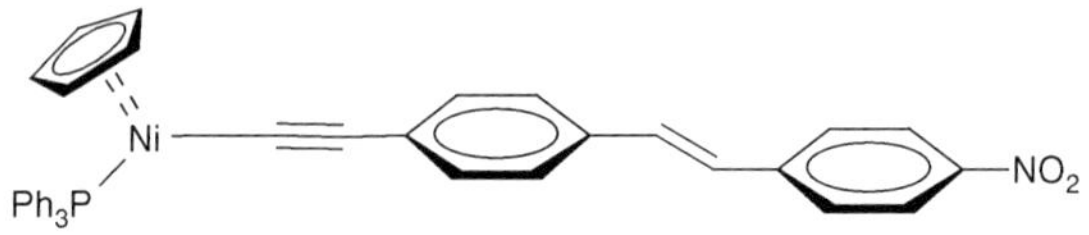

34

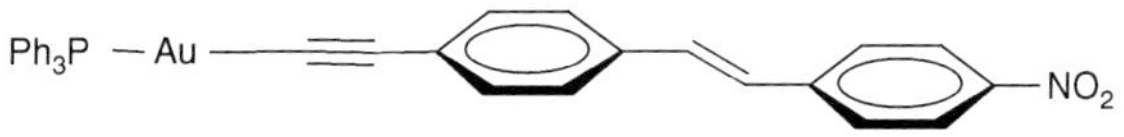

35

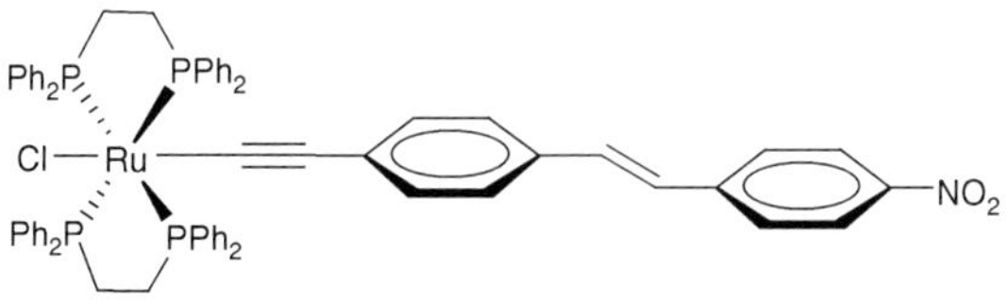

36

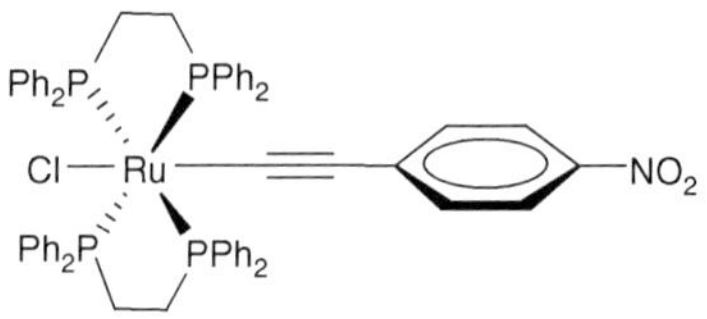

37

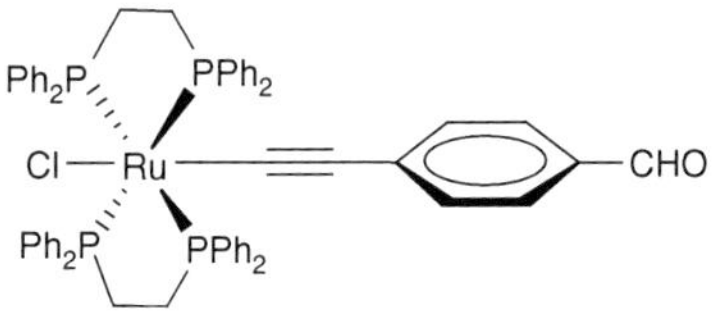

38

Analogous to metallocenes, in these complexes the metal acts as the donor group of a donor-acceptor system connected by a π-linker. The second-order nonlinearity is controlled by low-energy MLCT excitations. Large values of the quadratic hyperpolarizability, even if resonantly enhanced, have been achieved [11, 19, 144]. Accordingly, with the relative oxidizability ($Ru^{II} > Ni^{II} > Au^{I}$) [11, 19] the largest values of the quadratic hyperpolarizability have been obtained for the readily oxidizable Ru^{II} centers (e.g., **33**, $\beta_{1.06}$(HRS) = 1,455 × 10^{-30} esu), followed by the Ni^{II} complexes (e.g., **34**, $\beta_{1.06}$(HRS) = 445 × 10^{-30} esu), and by the 14-valence-electron, less readily oxidizable, Au^{I} species (e.g., **35**, $\beta_{1.06}$(HRS) = 120 × 10^{-30} esu) [11]. Group 8 complexes, in particular Ru^{II} compounds, consist of the largest group of acetylides studied [11, 19, 144–147], possessing the highest values of the quadratic hyperpolarizability (e.g., **36**, $\beta_{1.06}$(HRS) = 2,676 × 10^{-30} esu; β_0(HRS) = 342 × 10^{-30} esu) [145], thus resulting in very efficient donors. As expected, the quadratic hyperpolarizability [11, 19, 144–147] increases upon increasing the chain length of the acetylide bridge (e.g., **36** vs **37**, $\beta_{1.06}$(HRS) = 351 × 10^{-30} esu; β_0(HRS) = 55 × 10^{-30} esu, for the latter) [145], and increasing the strength of the acceptor (e.g., **37** vs **38**, $\beta_{1.06}$(HRS) = 120 × 10^{-30} esu, for the latter) [145]. Moreover, replacement of the cyclopentadienyl ligand by a tertiary-phosphine, always produces a substantial increase of the quadratic hyperpolarizability (e.g., **33** vs **36**). Compared to metallocene based second-order NLO chromophore, σ-acetylide based second-order NLO chromophores, which possess an almost linear M–C≡C–R structure, give rise to a better coupling between the d metal orbitals and the π* system of the σ-acetylide bridge and, hence, to larger second-order nonlinearity (e.g., **24**, $\beta_{1.91}$(EFISH) = 31 × 10^{-30} esu vs **37**). Bimetallic complexes involving σ-acetylide Ru^{II} complexes as donor and the "W(CO)₅" fragment as acceptor have also been investigated [19] (see Sect. 3.7).

Vinylidene [11, 145], aryldiazovinylidene [148], and alkenyl [149] species represent parallel families closely related to σ-acetylides. However, their second-order optical nonlinearity is generally lower [11]. Their interest is mostly related to the possibility of switching the NLO properties by virtue of the facile interconvertibility of acetylide and protonated vinylidene pairs via protonation/deprotonation sequences (e.g., **36** vs **39**, $\beta_{1.06}$(HRS) = 441 × 10^{-30} esu; β_0(HRS) = 74 × 10^{-30} esu, for the latter) [145] (Scheme 2).

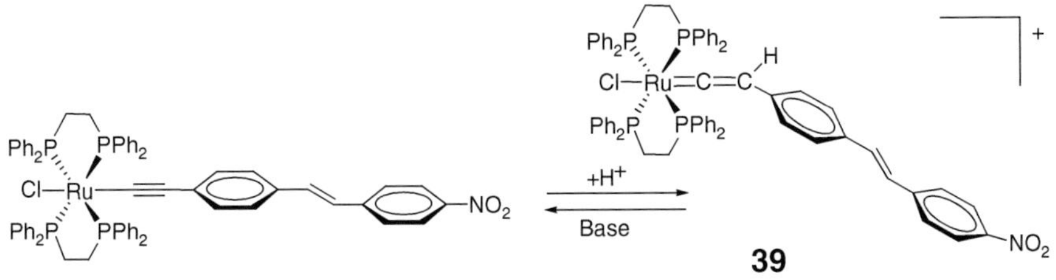

Scheme 2 Swtiching of second-order NLO response in the acetylide/vinylidene pairs upon protonation/deprotonation sequences

σ-Acetylide Ru^{II} complexes have also been investigated as octupolar and dendrimeric species [144, 150]. Actually, starting from the 1,3,5-substituted aromatic species, e.g., **40**, a variety of octupolar (e.g., **41**, $\sqrt{\langle\beta^2_{1.06}\rangle} = 1{,}220 \times 10^{-30}$ esu; $\sqrt{\langle\beta^2_0\rangle} = 254 \times 10^{-30}$ esu measured by HRS) or dendrimeric (e.g., **42**, $\sqrt{\langle\beta^2_{1.06}\rangle} = 1{,}880 \times 10^{-30}$ esu; $\sqrt{\langle\beta^2_0\rangle} = 350 \times 10^{-30}$ esu measured by HRS) structures have been synthesized and investigated [150]. They involve very large static and dynamic hyperpolarizabilities with an improved efficiency/transparency trade-off for SHG applications (e.g., **37**, $\lambda_{max} = 477$ nm vs **41**, $\lambda_{max} = 459$ nm).

Applications of these molecular NLO chromophores, to produce second-order bulk NLO materials or structured films, are very limited. Actually, in spite of their very large molecular hyperpolarizabilities, they exhibit crystalline materials with modest bulk SHG efficiency [11], e.g., the most efficient crystalline material shows a powder efficiency of only eight times that of urea [11], in consequence of the reluctance of acetylide complexes to crystallize in noncentrosymmetric structures.

Recently, poled polymer films involving, as guest, a dipolar Fe^{II} σ-acetylide NLO chromophore have shown the traditional temperature-dependent SHG properties [151], while a film of an Ru^{II} oligothienylacetylide NLO chromophore incorporated into a PMMA matrix revealed an acoustically induced SHG signal, reaching values ($\chi^{(2)} = 0.80$ pm V^{-1}) among the highest reported so far for this kind of unusual way to induce SHG [152].

In the last few years, this class of molecular materials has attracted renewed interest by virtue of their remarkable third-order NLO properties [11, 19, 150].

3.5 Cyclometallated Complexes

Recently, three classes of cyclometallated complexes have been reported to show interesting second-order NLO properties, therefore opening a novel route for the design of new efficient second-order NLO chromophores.

New orthopalladated NLO chromophores based on a parallel alignment of two different push-pull ligands have been synthesized by Centore et al. [153] and their second-order NLO activity has been measured by the EFISH technique working in $CHCl_3$ with an incident wavelength of 1.907 µm. The maximum value of $\mu\beta_{1.91}$(EFISH) (610 × 10^{-48} esu) was obtained for the NLO chromophore **43**.

43

Corona-poled thin polymeric films of **44** containing as guests this kind of NLO chromophores are characterized by large macroscopic NLO coefficients d_{33} (25 pm V^{-1}), as determined by means of SHG measurements working with an incident wavelength of 1.064 µm [154].

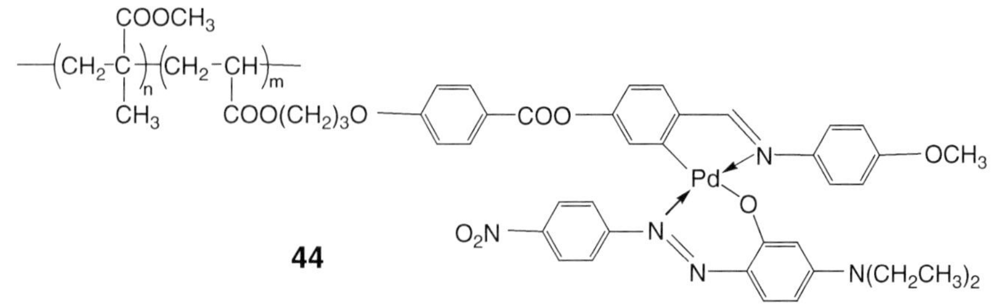

44

Recently, Labat et al. [155] studied the second-order NLO properties of a new cyclometallated Ru^{II} complex (**45**) for which the HRS technique, working in acetonitrile, gives a β_0(HRS) value of 230 × 10^{-30} esu. In this particular NLO chromophore the Ru^{II} moiety seems to act as the donor group of a push–pull system.

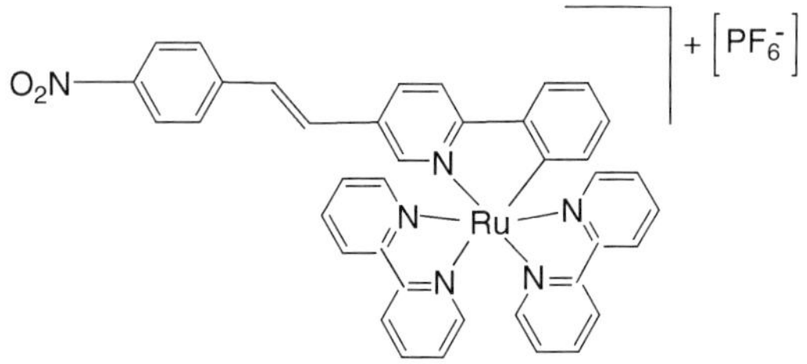

45

Finally, in the last 3 years, ionic cyclometallated Ir^{III} complexes with chelated π-delocalized ligands, such as bipyridines [156, 157] or phenanthrolines [158–160], have shown interesting photoemissive properties with potential applications in electroluminescent devices. Ugo et al. have extended their investigations on the second-order NLO response of these chromophores, such as for instance [Ir(cyclo-metallated-2-phenylpyridine)$_2$(5-R-1,10-phenanthroline)][PF$_6$] (R = H, Me, NMe$_2$, NO$_2$; **46**) and [Ir(cyclometallated-2-phenylpyridine)$_2$(4-R′,7-R′-1,10-phenanthroline)][PF$_6$] (R′ = Me, Ph; **47**) [160]. The quadratic hyperpolarizability was measured by the EFISH technique working with a nonresonant 1.907 μm incident λ in a low polarity solvent, like CH$_2$Cl$_2$, which allows by ion-pairing the extension of the use of this technique to ionic compounds. These Ir^{III} complexes show a large negative second-order response ($\mu\beta_{1.91}$(EFISH) ranging from −1,270 to −2,230 × 10^{-48} esu). Most notably, they do not show strong absorptions above 450 nm so that a significant SHG may be obtained without a significant cost in transparency. The highest absolute $\mu\beta_{1.91}$(EFISH) value is that of the complex carrying the phenanthroline with the strong electron-withdrawing group NO$_2$ [161]. A SOS-TDDFT theoretical investigation confirmed that the second-order NLO response of these Ir^{III} NLO chromophores is mainly controlled by the MLCT transition from the HOMOs phenylpyridine-Ir based orbital of the cyclometallated moiety to the

LUMOs π* orbitals of the phenanthroline [161]. Therefore, the second-order NLO response is strongly controlled by the donor or acceptor properties of the substituent on the phenanthroline ligand.

46   **47**

Substitution of cyclometallated 2-phenylpyridine with the more π-delocalized cyclometallated 2-phenylquinoline does not affect significantly the NLO responses, while a lower NLO response is obtained for the Ir^{III} NLO chromophore with a cyclometallated 3'-(2-pyridyl)-2,2':5',2"-terthiophene (ttpy), since the structure of ttpy induces a significant downshift of the HOMO's energy, compared to that of cyclometallated 2-phenylpyridine and 2-phenylquinoline [162].

3.6 Compounds with Macrocyclic Ligands

3.6.1 Metalloporphyrins

The macrocyclic structure of porphyrins, consisting of an extended π system formed by four pyrrolic rings connected by methine bridges, is a typical example of a very polarizable architecture with a variety of low lying excited states. Therefore the presence of various substituents in the *meso* or pyrrolic position of the porphyrin ring could produce significant perturbations. These structural features, together with the high chemical and thermal stability, can explain the widespread interest in these chromophores during the last few decades in the area of new optical materials.

Due to the high polarizability of the electronic cloud of the porphyrin ring, a large amount of work has been devoted to third-order NLO responses. For instance, we can refer to some comprehensive reviews [13, 163, 164] for applications based on third-order responses such as optical limiting (OL), for which metalloporphyrins are of great interest, given their tendency to show reverse saturable absorption (RSA) behavior because of their strong absorbing, long-lived triplet excited states and their transparency gap between the intense Soret (B) and Q π–π* absorption bands (in the range 400–500 nm and 600–700 nm, respectively). A certain amount of work has been devoted to two-photon absorption (TPA) responses of asymmetric porphyrinic structures [165]. The aim is not only that of possible applications in OL devices, requiring a high TPA absorption cross section, but also of exploiting their application in the field of photodynamic therapy [166]. In the last two decades a significant amount of work has also been devoted to the investigation of the second-order NLO properties of porphyrin architectures of increasing complexity; a recent review has been partially devoted to this specific area [164].

Taking into account the significant third-order NLO response of the porphyrin ring, when the asymmetric substitution of such a ring is too weak, the determination of the quadratic hyperpolarizability by the EFISH technique can be affected by significant errors because the third-order electronic contribution $\gamma(-2\omega; \omega, \omega, 0)$ to γ_{EFISH} cannot be neglected [167].

Pizzotti et al. reported [168] an EFISH investigation, working in $CHCl_3$ solution with a nonresonant incident wavelength of 1.907 μm, on the second-order NLO response of various push–pull tetraphenylporphyrins and their Zn^{II} complexes substituted at the β pyrrolic position by a π-delocalized organic substituent carrying either an electron-withdrawing or electron-donating group (**48**).

M = H$_2$, Zn; R = NBu$_2$, NMe$_2$, NO$_2$ **48**

Interestingly, the porphyrin ring shows in these push-pull NLO chromophores an ambivalent character as donor (due to the high polarizability of its electronic cloud) or as acceptor (due to the presence of various low-lying excited-states). When the substituent is an electron-acceptor group, the porphyrin ring behaves as a significant donor group, comparable to a ferrocenyl group. The value of $\beta_{1.91}$(EFISH) decreases only slightly on going from the free porphyrin to its Zn^{II} complex, in agreement with the assumption that the second-order NLO response is controlled by a CT process, favored by π conjugation, from the occupied π levels of the pyrrolic ring, acting as a push system, to the π* antibonding orbitals of the linker. This latter process should be scarcely affected by coordination of the porphyrin ring to Zn^{II}. In contrast, when the substituent is a strong electron-donor π-system, the $\beta_{1.91}$(EFISH) is not only higher but it increases by complexation to Zn^{II}, as expected for an increased acceptor property.

Diphenyl porphyrins and their Zn^{II} complexes substituted in the *meso* position by a π-delocalized substituent carrying an electron-donor or an electron-withdrawing group (**49**) were also investigated by the EFISH technique [169]. These second-order NLO chromophores have confirmed the ambivalent role of the polarizable porphyrin ring, which, in the ground state, already acts as a donor or acceptor, depending on the nature (acceptor or donor) of the substituent in the *meso* position. There is a significant increase of $\beta_{1.91}$(EFISH) for the same π-delocalized substituent carrying an electron-acceptor group, going from the substitution in the *meso* position to that in the β pyrrolic position. When the π-delocalized substituent is carrying an electron-donor group, the position of the substitution (*meso* or β pyrrolic) is influential on the value of $\beta_{1.91}$(EFISH). Interestingly, when the electron-acceptor substituent is in position *meso*, the donor property of the porphyrin ring becomes quite similar to that of the organic strong donor system *trans*-4-NMe$_2$C$_6$H$_4$CH=CH$_2$.

M = H$_2$, Zn; R = NBu$_2$, NMe$_2$, NO$_2$ **49**

A combined electrochemical, HRS and theoretical DFT investigation has also been carried out on *meso*-tetraphenylporphyrin (H$_2$TPP) and its first transition series metal complexes (MTPP) (**50**) [170].

While neutral MTPP, due to their centrosymmetric structure, have a zero second-order NLO response, the one and two-electron oxidized products of CuTPP and ZnTPP show significant β(HRS) values ($\beta_{1.06}$(HRS) = 351 and 371 × 10^{-30} esu, for CuTPP$^+$ and CuTPP^{++}; $\beta_{1.06}$(HRS) = 407 and 606 × 10^{-30} esu, for ZnTPP$^+$ and ZnTPP^{++}), confirmed by TD-DFT calculations. Electrochemical switching of their optical nonlinearity between the neutral and the oxidized forms has been achieved, but its repetition was demonstrated only for the first oxidation step.

50

M = H$_2$, Cr, Mn, Fe, Co, Ni, Cu, Zn

Axial coordination of stilbazoles like 4,4′-*trans* or *trans,trans*-Me₂N–C₆H₄(CH=CH)$_n$C₅H₄N (n = 1, 2) to tetraphenylporphyrinates of Zn^{II}, Ru^{II}, and Os^{II} [171] does not produce the increase of the quadratic hyperpolarizability which usually occurs when this kind of stilbazoles coordinates to hard or soft Lewis acid metallic centers (see Sect. 3.1.2). This lack of increase of the second-order NLO response upon axial coordination can be interpreted as being due to a significant axial π backdonation from the d$_π$ orbitals of the metal into the π* antibonding orbitals of the stilbazoles. This effect produces a contribution, opposite to that of σ donation, to the quadratic hyperpolarizability of the stilbazole, thereby giving rise to a balance of the positive (σ−donation) and negative (π-backdonation) effect on the quadratic hyperpolarizability. When 4,4′-*trans*-F₃C–C₆H₄(CH=CH)C₅H₄N is axially coordinated, the axial π backdonation becomes very relevant and prevails, thus resulting in a threefold increase of the EFISH quadratic hyperpolarizability. Therefore, the role of the axial π backbonding, when π-delocalized ligands, like stilbazoles, are axially coordinated to metal porphyrinates, causes the metal atoms of porphyrinates to act not only as σ acceptors, but also as π donors according to the nature of the stilbazoles.

From HRS measurements, working in CHCl₃ solution with a resonant incident wavelength of 1.064 μm, asymmetric *meso*-tetraaryl-metallo porphyrins such as **51** (Cu^{II}) and **52** (Zn^{II}) [172] have shown a relatively low second-order NLO response ($β_{1.06}$(HRS) = 118 and 92 × 10$^{−30}$ esu, respectively) when compared to asymmetric arylethynyl push–pull porphyniric NLO chromophores like **53** (Cu^{II}) and **54** (Zn^{II}) [173] ($β_{1.06}$(HRS) = 1,501 and 4,933 × 10$^{−30}$ esu, respectively), for which a very strong coupling between the donor and the acceptor substituents occurs [173]. In this latter case, the porphyrin ring acts mainly as a very polarizable and long π linker. The value of $β_{1.06}$ of **54**, deduced from both an absorption and electroabsorption investigation (Stark effect), has been subsequently reported to be lower ($β_{1.06}$ = 1,710 × 10$^{−30}$ esu), although still very high [174]. For **55**, carrying a Ni^{II} instead of a Zn^{II} or Cu^{II} metal center, much lower values have been measured by the EFISH technique working in CHCl₃ with a nonresonant incident wavelength of 1.907 μm [175]. Similar low values ($β_{1.91}$(EFISH) = 66–124 × 10$^{−30}$ esu) of the quadratic hyperpolarizability have been reported for structurally related porphyrin arylethynyl push–pull NLO chromophores measured by EFISH under the same experimental conditions [176]. Such striking differences have suggested that the metal could strongly influence the second-order NLO response of this kind of push-pull NLO chromophores. However, recent theoretical DFT or HF and coupled-perturbed (CP) DFT or HF investigations on the linear and second-order nonlinear properties of **53**, **54** and **55** [177] have shown that their second-order NLO response is barely affected by changing the metal. Moreover, the values of $β_0$ of **54**–**55** calculated by CP-HF and CP-DFT level of theory are similar and in the range 61–66 × 10$^{−30}$ esu and 301–327 × 10$^{−30}$ esu, respectively, with $β_{1.91}$(CP-HF) values in the range 76–79 × 10$^{−30}$ esu. These are much lower values than those reported for **54**, when measured by HRS working at a resonant incident wavelength of 1.064 μm. The discrepancy of the experimental measurements is probably due to the different incident wavelengths, as confirmed by the comparison of the calculated $\beta_{1.06}$ and $\beta_{1.91}$ values of **54** at HF level of theory, with the former showing a much higher value, as a result of a resonance between the second harmonic (532 nm) and the strong Q band at about 600 nm.

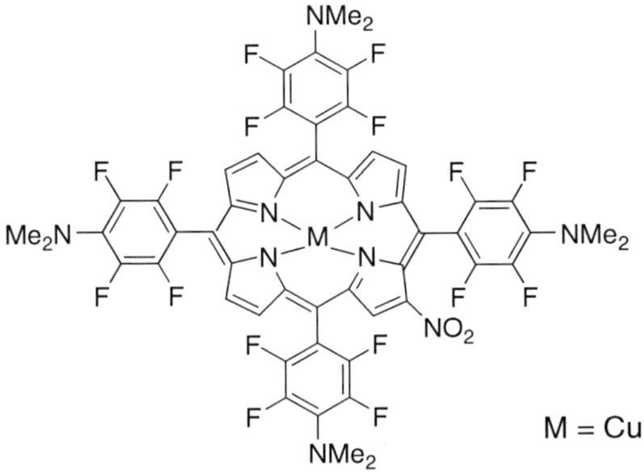

M = Cu (**51**), Zn (**52**)

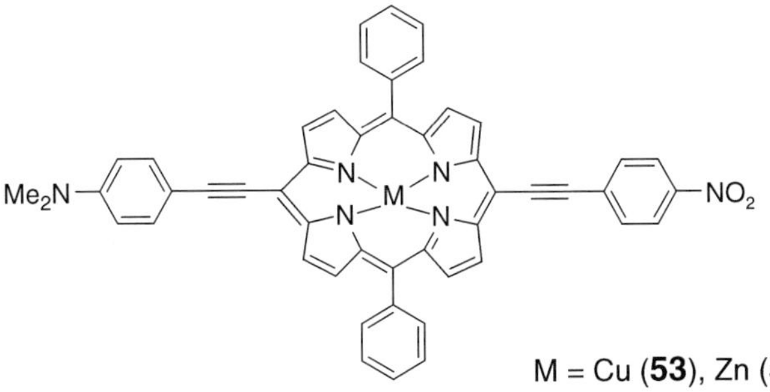

M = Cu (**53**), Zn (**54**), Ni (**55**)

An attempt to link covalently an NLO chromophore structurally related to **54** to a polymeric network has been made by introducing a methacrylate group on the donor part of the molecule and a carboxylic acid function on the acceptor one (**56**) [178]. Copolymerization with glycidyl methacrylate has been successfully carried out, affording a composite polymeric film whose electrooptical properties are under investigation.

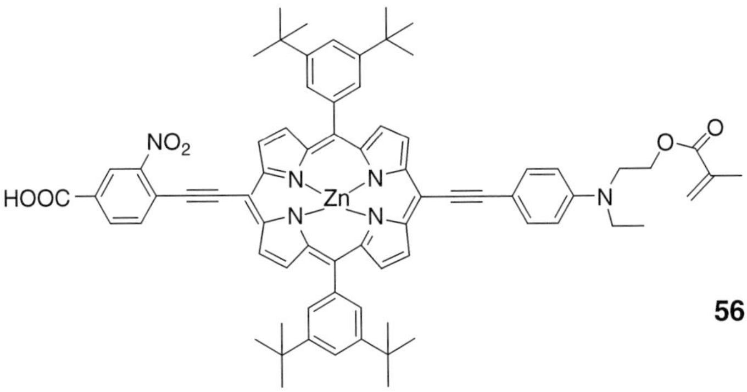

56

In order to improve the second-order NLO response, working at 1.30 μm incident wavelength, new push-pull NLO chromophores based on the porphyrin ring have been synthesized, coupling to the porphyrin ring thiophene or thiazole rings.

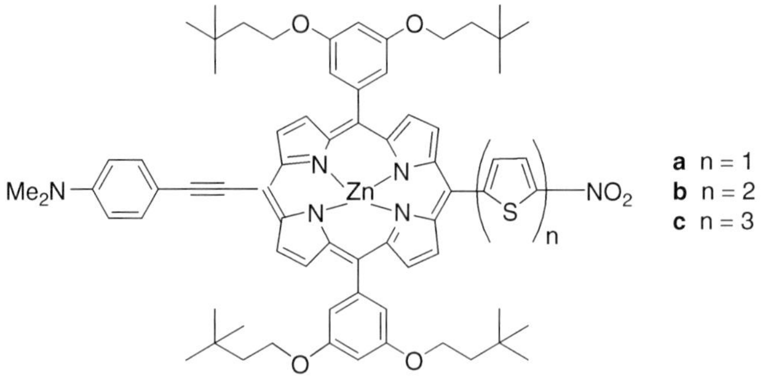

57

The second-order NLO response of molecular architectures such as **57** and **58** [179] or **59** and **60** [180] (structurally related to **58a**) has been determined by the HRS technique in THF solution. $\beta_{1.30}$(HRS) values from 650 to 4,350 × 10^{-30} esu have been measured for **57** and **58** (the highest value was obtained for **57c**). The lower values of the quadratic hyperpolarizability have been measured when thiophene or oligothiophene units are linked to the porphyrinic core through a C≡C triple bond (**58**). For the NLO chromophores **59** and **60**, $\beta_{1.30}$(HRS) values of about 785–1,400 × 10^{-30} esu have been measured, with **60b** showing the highest value. It must be pointed out, however, that the second harmonic at 0.65 μm (650 nm) is in the region of intense Q absorption bands. Therefore, it is possible that the values of the quadratic hyperpolarizabilities could be affected by a significant enhancement due to resonance effects. A detailed theoretical analysis of the origin of the quadratic hyperpolarizability of these NLO chromophores, which is beyond the scope of this review, can be found in [179] and [180]. In conclusion, chromophores such as **57**–**60** may be interesting for long-wavelength optoelectronic applications.

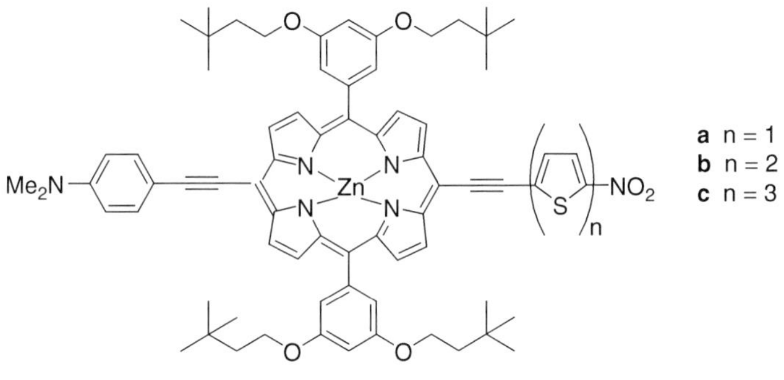

58

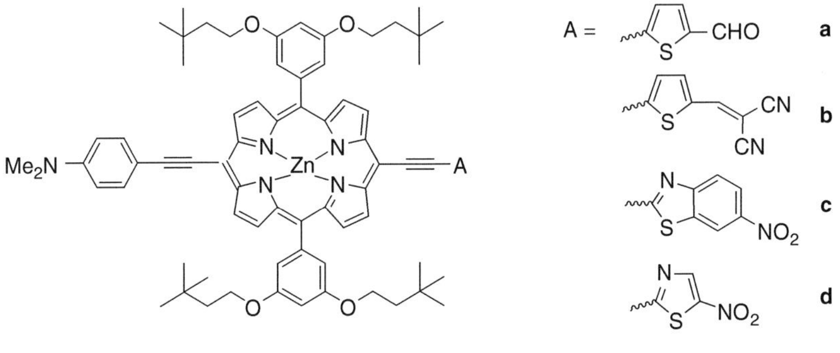

59

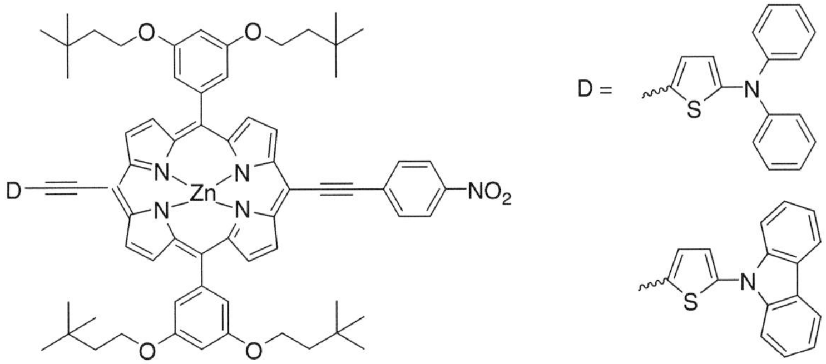

60

The effect of the central metal ion on the value of the quadratic hyperpolarizability has been studied for chromophores similar to **58b**, but carrying in the phenyl ring in 5,15 positions a CH_3 instead of a 3,5-bis(3,3-dimethyl-1-butyloxy)phenyl substituent. Metals such as Mg, Co, Ni, Cu, and Zn have been considered [181]. Their quadratic hyperpolarizabilty was theoretically evaluated by semiempirical ZINDO/CV calculations, which have shown that metalloporphyrins, due to significant CT transitions, may display values of the quadratic hyperpolarizability about one order of magnitude higher than that of the free porphyrins. Moreover, by varying the metal atom, an increase of the quadratic hyperpolarizability could be achieved, the highest value being computed for the NLO chromophore with Mg as metal center ($\beta_{1.91} = 1{,}120 \times 10^{-30}$ esu).

In order to investigate how the second-order NLO properties of this kind of NLO chromophores are affected by aggregation of many porphyrin rings, push-pull metalloporphyrins carrying two or three porphyrin rings, such as those reported below (**61**) [182] where the single porphyrin rings are connected by two C≡C triple bonds, have been theoretically investigated.

M = Mg, Co, Ni, Cu, Zn
R = CH₃   **61**

Semiempirical ZINDO/CV calculations have suggested that the quadratic hyperpolarizability of dimers and trimers are about one order of magnitude higher than that of the monomer [182]. A DFT/TDDFT investigation on the NLO chromophores **57b** and **58b** [183] has confirmed an almost planar architecture, which enhances the π-conjugation of the push–pull system and, as a consequence, the quadratic hyperpolarizability.

Structures with a Zn^{II} porphyrin cycle and a metal^{II}polypyridyl connected through the 10,20 *meso* position of the porphyrin by a C≡C linker have been studied by HRS working in CH$_3$CN solution with a 0.80 μm incident wavelength (**62**) [184]. HRS depolarization experiments have shown that the second-order NLO response of these architectures is mainly due to conformers in which the torsional angles between the polypyridyl unit and the porphyrin core are opposite in sign but equivalent in magnitude (θ ≈ −ϕ). These species can thus be considered as interesting building blocks for nonpolar chiral electrooptic materials.

R = C$_3$F$_7$, M = Ru, Os
R = 2,6-bis(3,3-dimethylbutyloxy)phenyl, M = Os

3.6.2 Metallophthalocyanines

Phthalocyanines are macrocycles characterized by an extensive 2D planar and centrosymmetric 18 π-electron system. For this reason they have been widely investigated as third-order NLO materials and in particular as potential materials for OL; some recent reviews have appeared in this specific field [163, 185, 186]. Only in the last decade have the second-order NLO properties been investigated and some of the reviews reported above describe the first significant results [164, 185]. Hereafter, we will highlight the must relevant results achieved since 2003 in the field of phthalocyanines showing second-order NLO properties.

By means of the EFISH technique, working in CHCl$_3$ with a nonresonant incident wavelength of 1.907 μm, the second-order NLO response of the asymmetric phthalocyanine **63** has been measured and compared to that of phthalocyanines **64** and **65** [187].

The dipole moments of push-pull NLO chromophores **63** and **64a** are exceptionally high (38.8 D and 33.6 D, respectively), while the measured γ(EFISH) values are negative for all of them (γ(EFISH) = -41.3×10^{-34} esu for **64a**;

γ(EFISH) = -57.4×10^{-34} esu for **64b**; γ(EFISH) = -16.8×10^{-34} esu for **65**) except for **63** (γ(EFISH) = 5.25×10^{-34} esu). The high value found for the centrosymmetric phthalocyanine **64b** clearly shows that, for this kind of second-order NLO chromophores, the electronic contribution γ ($-2\omega; \omega, \omega, 0$) to the EFISH measurement is significant and cannot be neglected. The comparison between the positive value obtained for **63** and the negative value measured for **64a**, which bears a similar substitution pattern but without a triazolehemiporphyrazine bridge between the donor and the acceptor part of the molecule, suggests a change in the sign of $\Delta\mu_{eg}$.

In a further investigation [188], a new family of substituted push-pull phthalocyanines (**66** and **67**), carrying triple bonds as linkers between the donor and the acceptor groups, has been investigated.

The second-order NLO response of these push–pull NLO chromophores has been measured by means of both EFISH (in CHCl$_3$, working at 1.064 and 1.907 µm incident wavelength) and HRS (working at 1.064 µm incident wavelength) techniques. The largest $\beta_{1.91}$(EFISH) and $\beta_{1.91}$(HRS) values (522×10^{-30} esu and 530×10^{-30} esu, respectively) were obtained for **66b**, which exhibits the highest degree of dipolar asymmetry. The ethynyl based linker seems, therefore, to be an excellent spacer for enhancing the second-order NLO response of this kind of asymmetric push-pull phthalocyanines.

Coordination and Organometallic Complexes

In order to overcome the nonlinearity transparence trade-off (the lengthening of the π conjugation increases the second-order NLO response of this kind of second-order NLO chromophores, but at the same time decreases their transparency), a bisphthalocyanine, with the CT between the donor and the acceptor guaranteed through space by a [2,2] paracyclophane unit (**68**), has been synthesized [189] and its second-order NLO response measured in CHCl$_3$ by both EFISH and HRS, working at 1.064 μm incident wavelength; a significant value of $\beta_{1.06}$(HRS) (180 × 10^{-30} esu) was measured.

Of particular interest for their high second-order NLO response are the intrinsically noncentrosymmetric phthalocyanines analogs called subphthalocyanines. These cone-shaped macrocycles, consisting of three isoindole moieties with a central boron atom coordinated to an axial halogen ligand, have been studied for both their dipolar and their octupolar characters, given that they possess not only a dipole moment along the B-halogen bond, but also an octupolar charge distribution within their three-dimensional aromatic architecture [185]. For instance, the NLO chromophore **69** is characterized by a significant $\beta_{1.34}$(HRS) value (104 × 10^{-30} esu), due to the strong octupolar character of the subphthalocyanine core and to a fair $\gamma_{1.34}$(EFISH) value (17.9 × 10^{-34} esu) corresponding mainly to the dipolar orientational contribution to $\gamma_{1.34}$(EFISH). A fine-tuning of the dipolar and octupolar contributions was reached by varying the substituents of the subphthalocyanine core [190].

The effect of a static electric field on the SHG of the centrosymmetric copper phthalocyanines **70** has been investigated, applying an external d.c. voltage to an Au-phthalocyanine film-Au system that induces a second-order polarization [191]. A SHG signal has also been obtained at the interface of a Langmuir-Blodgett film of **70b**, deposited on a metal-coated glass slide [192] and from a film obtained by vacuum-evaporation of **70a** at a metal electrode interface [193].

3.7 Bimetallic Complexes

The investigation of the second-order NLO response of asymmetric bimetallic complexes in which the electron-accepting and donating properties of two metal-based fragments are combined has been an area of interest in the past few years [13, 19].

The ferrocenyl moiety has a donor strength comparable to that of a methoxyphenyl group [194]; therefore many bimetallic complexes containing a ferrocenyl moiety linked via a π-delocalized bridge to another organometallic fragment, acting as acceptor group, have been investigated [13, 19]. For example, when η^7-cycloheptatrienyltricarbonylchromium is used as acceptor group, a high quadratic hyperpolarizability has been measured by HRS (**71**, $\beta_{1.06}$(HRS) = 570 × 10^{-30} esu) [195].

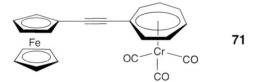

The "W(CO)$_5$" fragment, which can behave as donor or acceptor in monometallic NLO chromophores (see Sect. 3.1), appeared as an efficient acceptor group in these NLO chromophores with ferrocene as donor. For example, the $\beta_{1.06}$(HRS) value of *trans*-(1-ferrocenyl-2-(4-pyridyl)ethylene] increases by a factor of about 5 upon coordination to "W(CO)$_5$" (**72**, $\beta_{1.06}$(HRS) = 101 × 10^{-30} esu). The EF is less significant when changing the metal and it decreases in the order "W(CO)$_5$" > "Mo(CO)$_5$" > "Cr(CO)$_5$," probably due to the higher π-backdonation of the electron *d* density in the case of W, if compared to Cr and Mo [196]. Interestingly, no significant EF is observed upon coordination of the pyridyl ligand to "Re(CO)$_3$Br" [197]. As expected, an increase in the conjugation length of the linker, for

Coordination and Organometallic Complexes

instance by insertion of a vinylenephenylene between the ethylene and the ferrocenyl, results in higher $\beta_{1.06}$(HRS) values (for *trans,trans*-(η^5-C$_5$H$_5$)Fe {η^5-C$_5$H$_4$CH=CHC$_6$H$_4$CH=CH–C$_5$H$_4$N–W(CO)$_5$, 535 × 10^{-30} esu [120]). Coordination of *trans*-[1-ferrocenyl-2-(4-cyanophenyl)ethylene] to "W(CO)$_5$" (**73**) also leads to an enhancement of the $\beta_{1.06}$(HRS) value (by a factor of 3.6; $\beta_{1.06}$(HRS) = 375 × 10^{-30} esu), less significant than that of the related ferrocenyl system based on the donor pyridine ligand (**72**) [196].

72

73

Various bimetallic complexes with ferrocenyl as electron-donor group linked via a π-delocalized system and pyridine coordination to electron-deficient "Mo^{II}(NO)Tp*X" (Tp* = tris(3,5-dimethylpyrazolyl)borate; X = Cl, Br, I) or "W(CO)$_5$" centers have been studied by McCleverty et al. [132]. For example, the complex **74** is characterized by a $\beta_{1.06}$(HRS) value of 205 × 10^{-30} esu. Although the measured HRS data are resonance enhanced, the $\beta_{1.06}$(HRS) trends imply the following conclusions: (1) the quadratic hyperpolarizability increases when the ligand Cl$^-$ or Br$^-$ is replaced by I$^-$, which is reasonable in the light of the more substantial polarizability of I$^-$ when compared with Cl$^-$ and Br$^-$; (2) methylation of ferrocene increases its electron-donor ability; and (3) the electron-accepting abilities of the "Mo^{II}(NO)Tp*Cl" and "W(CO)$_5$" moieties are similar. Furthermore, chemical oxidation of the ferrocenyl donor group or chemical reduction of the molybdenum nitrosyl acceptor fragment in selected NLO chromophores caused a reduction of between 25% and 100% of the NLO response, therefore allowing redox-induced switching of the NLO responses [132].

Jayaprakash et al. [198] reported the series of push–pull bimetallic polyene complexes [(CO)$_5$M=C(OCH$_3$)(–CH=CH–)$_n$(η^5C$_5$H$_4$)Fe(η^5C$_5$H$_5$)] (M = W, Cr; n = 1–4) with ferrocene as donor and a carbene complex of Cr or W as acceptor. The $\beta_{1.06}$(HRS) values increase with increasing the conjugation of the π linker; for the same π linker, the chromophore based on W exhibits higher second-order NLO responses than that based on Cr, as expected for a more efficient π backdonation from the W atom. The quadratic hyperpolarizability of these bimetallic push–pull polyene NLO chromophores exhibits significant solvent polarity dependence, suggesting a CT from the ferrocene donor to the carbene acceptor along the polyene backbone with significant $\Delta\mu_{eg}$. The highest $\beta_{1.06}$(HRS) value (780 and 2,420 × 10^{-30} esu, in hexane and acetonitrile, respectively) was reached for the NLO chromophore with M = W and n = 4.

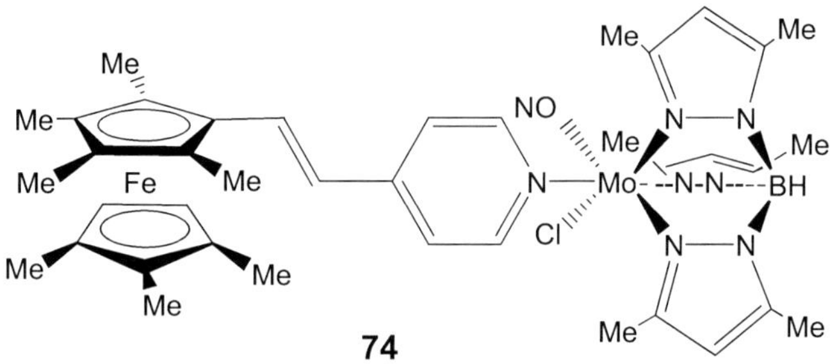

74

In this kind of bimetallic chromophores, another efficient electron-donor group is a ruthenium σ-acetylide complex [13, 19] which can lead to large $\beta_{1.06}$(HRS) values (535 and 700 × 10^{-30} esu for **75** and **76**, respectively [199]). Comparison with complexes **72** and **73** confirms that a ruthenium σ-acetylide complex is a better donor than the 1-ferrocenyl-ethylene moiety. In all these NLO chromophores, "W(CO)$_5$" is an efficient acceptor group, better than "Cr(CO)$_5$" [199].

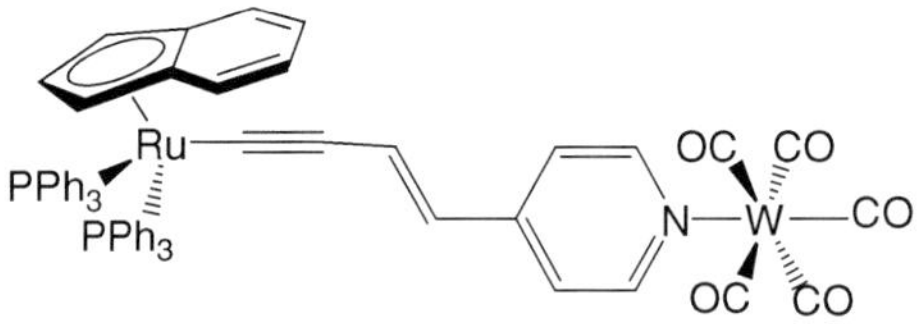

75

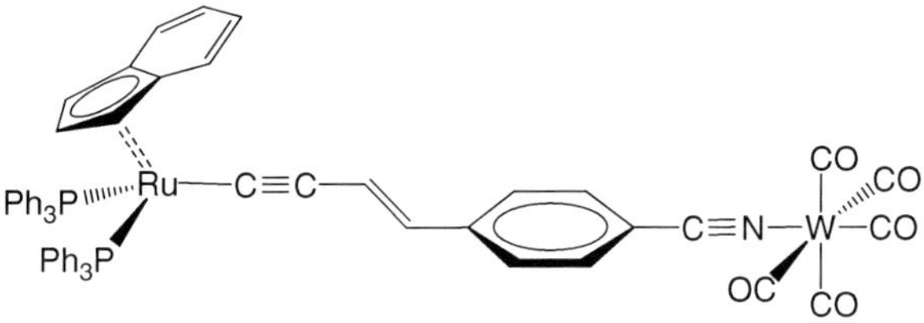

76

Significant quadratic hyperpolarizabilities for two mixed-valence bimetallic complexes have also been reported [13, 19]; for example complex **77** is characterized by a $\beta_{1.06}$(Stark) of 37 × 10^{-30} esu [200]. In such NLO chromophores, the electron-accepting d^5 Ru^{III} center is stabilized by strongly basic amine ligands, whilst the electron-donating d^6 Ru^{II} is stabilized by π-accepting cyanides.

$$\left[\begin{array}{c} \mathrm{H_3N} \mathrm{NH_3} \mathrm{NC} \mathrm{CN} \\ \mathrm{H_3N} - \mathrm{Ru} - \mathrm{N} \equiv \mathrm{C} - \mathrm{Ru} - \mathrm{CN} \\ \mathrm{H_3N} \mathrm{NH_3} \mathrm{NC} \mathrm{CN} \end{array} \right] \mathrm{Li} \cdot 3\mathrm{H_2O}$$

77

The second-order NLO response of a class of asymmetric push–pull bimetallic complexes, organometallic counterpart of classical push–pull aromatic chromophores, having pyrazine (pyz) and *trans*-1,2-bis(4-pyridyl)ethylene (bpe) as π-polarizable linkers between a metal carbonyl fragment acting as donor group ("W(CO)$_5$") and a metal carbonyl fragment acting as acceptor group (e.g., "*cis*-Rh(CO)$_2$Cl"), was investigated [201]. Coordination of (CO)$_5$W(pyz, or bpe) to "*cis*-Rh(CO)$_2$Cl" leads to an increase of the absolute value of the quadratic hyperpolarizability ($\beta_{1.91}$(EFISH) = −33 and −41 × 10^{-30} esu, for **78** and **79**, respectively), dominated by a red-shift of the MLCT (W^0→π*) transition of the W(CO)$_5$(pyz or bpe) moiety, due to the stabilization by interaction with the "*cis*-Rh(CO)$_2$Cl" acceptor group of the π* acceptor orbitals of the pyz or bpe bridge [201].

(CO)$_5$WN⌬N*cis*-Rh(CO)$_2$Cl

78

(CO)$_5$WN⌬=⌬N*cis*-Rh(CO)$_2$Cl

79

The CT is in the direction opposite to that of the ground state dipole moment, leading to a negative quadratic hyperpolarizability. In these NLO chromophores, the MLCT process (W^0→π*), dominating the NLO response, remains localized on the part of the π bridge close to the push donor center ("W(CO)$_5$") even after coordination at the other end of the bridge of a metallic center acting as pull acceptor center. Therefore, the electronic process controlling the second-order NLO response never assumes a MMCT (metal-to-metal CT) character, with the CT process involving both the donor and acceptor metal centers via the polarizable π bridge as it occurs between the donor and acceptor groups in structurally related classical push-pull aromatic 1D chromophores. In accordance, in these bimetallic NLO chromophores the quadratic hyperpolarizability is not strongly influenced by increasing the length of the π-delocalized bridge between the push and pull groups, while in structurally related organic π-delocalized push–pull 1D chromophores the second-order NLO response increases by increasing the length of the π-delocalized bridge [60, 61]. Consequently the value of $\beta_{1.91}$(EFISH) of **78** is, as absolute value, higher than that of *para*-nitro dimethyl aniline (12×10^{-30} esu), whereas that of **79** is lower than that of *trans*-4-dimethylamino, 4-nitro stilbene (73×10^{-30} esu) [201]. Theoretical TD-DFT and TD-HF calculations have confirmed such a view of the origin of the second-order NLO response in these push–pull NLO chromophores [202].

An interesting effect due to metal–metal bonding was studied in the case of bimetallic NLO chromophores such as [(CO)$_3$(1,10-phenanthroline)W-*fac*-MCO)$_3$Cl$_2$] (M = Ru^{II}, Os^{II}) and [(CO)$_3$(1,10-phenanthroline)W-*cis*-MCO)$_2$Cl] (M = Rh^I, Ir^I) which show an increased negative second-order NLO response ($\beta_{1.91}$(EFISH) from -25.8×10^{-30} up to -76×10^{-30} esu), the best value being measured for the Os^{II} derivative. The increase of the absolute value of the second-order NLO response when compared to that of the monometallic NLO chromophore [W(CO)$_4$(1,10-phenanthroline)] ($\beta_{1.91}$(EFISH) = -13×10^{-30} esu) is due, as shown by a solvatochromic investigation, to the additional negative contribution produced by a new absorption band involving the W(CO)$_3$(1,10-phenanthroline) moiety at around 385–450 nm, in addition to the contribution of the MLCT (W$^0 \rightarrow \pi^*$phen) absorption band at around 499–526 nm, typical of [W(CO)$_3$(1,10-phenanthroline)L] (L = CO, CH$_3$CN) [203].

Recently Coe et al. synthesized some new push–pull bimetallic complexes with *trans*-[Ru^{II}Cl(pdma)$_2$]$^+$ (pdma = 1,2-phenylenebis(dimethylarsine)) as electron-donor group, linked via a π-conjugated bridging ligand to *fac*-[Re^I(biq)(CO)$_3$]$^+$ (biq = 2,2′-biquinolinyl) as electron-acceptor group [204]. The bridging units were 4,4′-bipyridyl (4,4′-bpy; **80**), *trans*-1,2-bis(4-pyridyl)ethylene (bpe; **81**), or 1,4-bis[*trans*-2-(4-pyridyl)ethenyl]benzene (bpvb; **82**). The electronic absorption spectra of these bimetallic species are dominated by intense and opposite Ru$^{II} \rightarrow \pi^*$(4,4′-bpy/bpe/bpvb) and Re$^I \rightarrow \pi^*$(biq) MLCT transitions. Cyclic voltammetric studies reveal both Ru$^{III/II}$ oxidation and ligand-based reduction processes and show no evidence for significant electronic communication between the two metal centers. Stark spectroscopic studies on the MLCT bands show that extending the π conjugation of the linker leads to an increase of the dipole moment change ($\Delta\mu_{eg}$) and of the transition dipole moment (r_{eg}). These effects combine to afford an increase of the static quadratic hyperpolarizabilities, β_0, estimated by applying the "two level" model (β_0(Stark) = 47, 114 and 122×10^{-30} esu, for **80**, **81**, and **82**, respectively). Comparisons with monometallic Ru^{II} complexes reveal that methylation of the free pyridyl nitrogen, to generate a pyridinium center, leads to larger β_0 responses (β_0(Stark) = 113, 123, and 131×10^{-30} esu, for the alkylated monometallic Ru^{II} complex with 4,4′-bpy, bpe, or bpvb, respectively) than does coordination to the *fac*-[Re^I(biq)(CO)$_3$]$^+$ center. The outcome can be attributed to the evidence that the Re^I moiety is a weaker net electron-acceptor. In fact, it is a Lewis acceptor but it also behaves as a π donor, and this offsets partially its electron-withdrawing effect. In contrast, an *N*-methylpyridinium group does not possess such an ambivalent electronic behavior [204].

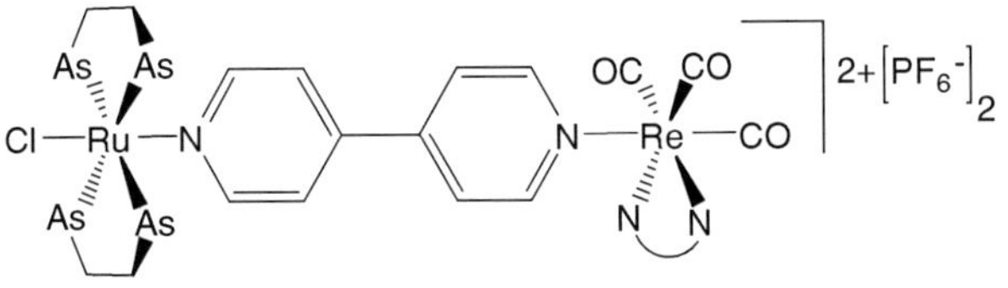

80

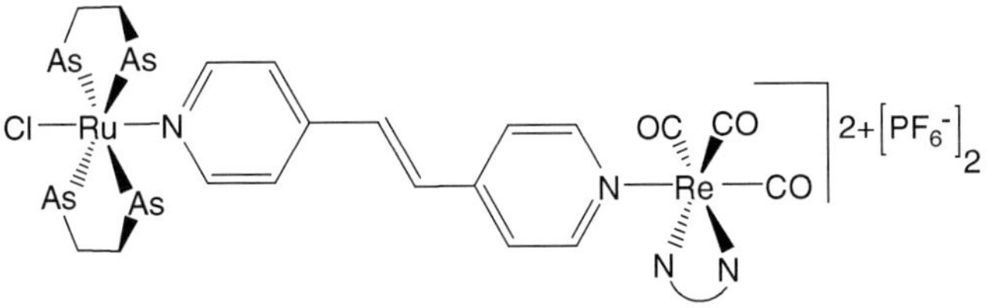

81

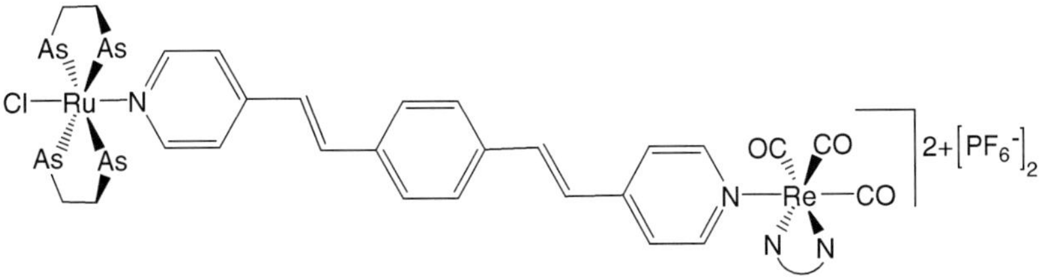

82

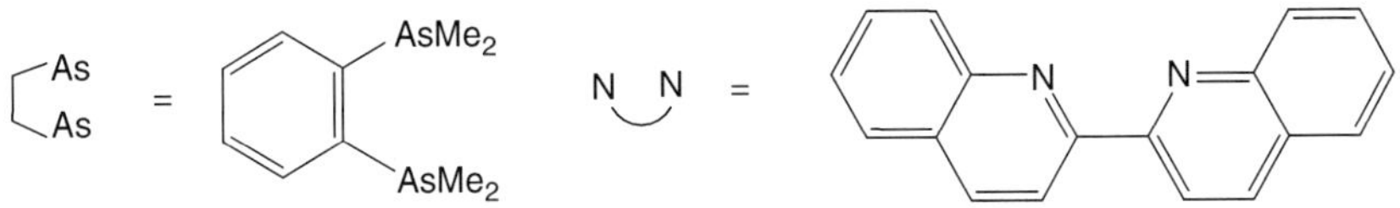

As evidenced from these examples, the introduction of two metal centers as donor and acceptor groups in a push–pull NLO chromophore can allow tuning of the electronic properties for a better second-order NLO response, affording relatively large quadratic hyperpolarizabilities.

4 Conclusions and Perspectives

This short and not exhaustive overview illustrates the actual level of the control of the structure – second-order NLO activity relationship in organometallic or coordination compounds and therefore their potentiality as second-order NLO chromophores.

Many coordination or organometallic compounds with various ligands such as substituted stilbazoles, bipyridines, phenanthrolines, terpyridines, Schiff bases, alkynyl and vinylidene ligands, metallocenes, or macrocycles such as metalloporphyrins, metallophthalocyanines, or the new emerging class of cationic cyclometalated or of bimetallic coordination complexes, have been investigated as second-order NLO chromophores in the last 10 years. They are characterized by rather large β_λ(HRS), $\mu\beta_\lambda$(EFISH), or $\mu\beta_0$(EFISH) values which can be favorably compared with that of the prototypical organic 1D push-pull NLO chromophore Disperse Red One (*trans*-4,4'-O$_2$NC$_6$H$_4$N=NC$_6$H$_4$NEt–(CH$_2$CH$_2$OH), $\mu\beta_0$(EFISH) = 450 × 10^{-48} esu) currently proposed for electrooptic polymeric poled films [205, 206]. Moreover, a variety of push-pull Ru^{II} ammine salts and of octupolar metal (in particular of Ru^{II}) complexes, recently studied, are characterized by large static quadratic hyperpolarizabilities, as determined by HRS.

It appears from this overview that the interaction with a metal center represents a flexible way to tune the second-order NLO response of organic NLO chromophores acting as ligands. By the modulation of the electronic nature and the oxidation state of the metal and of its coordination sphere through ancillary ligands or chelation, such a response can not only be increased, sometimes in a very significant way, but its sign can also be changed. Besides, the metal may act as template for the stabilization of octupolar D_3 and D_{2d} structures, based on various chelated organic ligands, or of asymmetric Schiff-bases complexes, producing second-order NLO chromophores, either octupolar or dipolar, with significant and tunable second-order NLO response. Finally, it is possible to produce bimetallic push-pull NLO chromophores where metal based moieties, acting as donor or acceptor groups, are connected by a π organic linker. In these cases, there are significant differences with the corresponding traditional push–pull organic NLO chromophores about the general trends and the origin of the quadratic hyperpolarizability. The metal can also act as a bridge, thus permitting electron-transfer processes between various sites of its coordination sphere such as in some push–pull metal porphyrins or Schiff bases complexes and in the new cyclometallated chromophores. The great potentiality of TD-DFT, CP-DFT, and CP-HF theoretical approaches has recently given a way to understand better the electronic origin of the second-order NLO response of many NLO chromophores based on metal complexes or organometallic compounds. These useful theoretical tools may allow today the design of new and efficient coordination and organometallic second-order NLO chromophores.

Sufficiently large second-order NLO responses have already been achieved; therefore the research now hinges also on considerations such as thermal and chemical stability and multifunctionality (for example, combining second-order NLO response with emission or magnetic properties). Nanoorganization of second-order NLO chromophores based on metal complexes or organometallic compounds, for example in Langmuir–Blodgett films or in polymeric electrically poled films, has been investigated but is still under study in order to reach large and stable in time SHG. In the last few years, increasing effort has also been dedicated to achieve the switching of the second-order NLO response, for example by redox-induced switching or photo-switching controlled by the metal.

Therefore, metal complexes and organometallic compounds are attractive not only for their interesting second-order NLO properties but also for their potentiality as multifunctional materials. Up to now, however, no stable and efficient nanostructured materials or electro-optical devices based on coordination and organometallic compounds have reached the stage of real applications and commercialization. But this is normal for a relatively young field of research, of which almost all scientific contributions have appeared in the last two decades. In any case, given the rapid progress made over recent years, future prospects for applications seem possible.

Acknowledgements We sincerely thank Dr Lea Vaiana for assistance in drawing various figures, the Ministero dell'Istruzione, dell'Università e della Ricerca (Progetto FIRB 2003 RBNE03-3KMA Molecular compounds and hybrid nanostructured materials with resonant and non resonant optical properties for photonic devices) and the Centro Nazionale delle Ricerche (PROMO 2006 Nanostrutture organiche, organometalliche, polimeriche ed ibride: ingegnerizzazione supramolecolare delle proprietà fotoniche dispositivistiche innovative per optoelettronica) for financial support.

References

1. Prasad NP, Williams DJ (1991) Introduction to nonlinear optical effects in molecules and polymers. Wiley, New York
2. Zyss J (1994) Molecular nonlinear optics: materials, physics and devices. Academic, Boston
3. Roundhill DM, Fackler JP Jr (eds) (1999) Optoelectronic properties of inorganic compounds. Plenum, New York
4. Green MLH, Marder SR, Thompson ME, Bandy JA, Bloor D, Kolinsky PV, Jones RJ (1987) Nature 330:360–362
5. See for example Nalwa HS (1991) Appl Organomet Chem 5:349–377
6. Marder SR (1992) In: Bruce DW, O'Hare D (eds) Inorganic materials. Wiley, New York, pp 115–164
7. Long NJ (1995) Angew Chem Int Ed Engl 34:21–38
8. Whittall IR, McDonagh AM, Humphrey MG, Samoc M (1998) Adv Organomet Chem 42:291–362
9. Heck J, Dabek S, Meyer-Friedrichsen T, Wong H (1999) Coord Chem Rev 190/192:1217–1254
10. Le Bozec H, Renouard T (2000) Eur J Inorg Chem 229–239
11. Powell CE, Humphrey MG (2004) Coord Chem Rev 248:725–756
12. Di Bella S (2001) Chem Soc Rev 30:355–366
13. Coe BJ (2004) In: McCleverty JA, Meyer TJ (eds) Comprehensive coordination chemistry II. Elsevier, Oxford, pp 621–687
14. Coe BJ, Curati NRM (2004) Comments Inorg Chem 25:147–184
15. Maury O, Le Bozec H (2005) Acc Chem Res 38:691–704
16. Cariati E, Pizzotti M, Roberto D, Tessore F, Ugo R (2006) Coord Chem Rev 250:1210–1233
17. Coe BJ (2006) Acc Chem Res 39:383–393
18. Coe BJ (2006) In: Papadopoulos MG (eds) Non-linear optical properties of matter. Springer, Berlin Heidelberg New York, pp 571–608

19. Humphrey MG, Samoc M (2008) Adv Organomet Chem 55:61–136
20. Oudar JL, Chemla DS (1977) J Chem Phys 66:2664–2668
21. Oudar JL (1977) J Chem Phys 67:446–457
22. Kanis DR, Ratner MA, Marks TJ (1994) Chem Rev 94:195–242
23. Marques MAL, Gross EKU (2004) Annu Rev Phys Chem 55:427–455
24. Ledoux I, Zyss J (1982) Chem Phys 73:203–213
25. Maker PD (1970) Phys Rev A 1:923–951
26. Clays K, Persoons A (1991) Phys Rev Lett 66:2980–2983
27. Zyss J, Ledoux I (1994) Chem Rev 94:77–105
28. Bruni S, Cariati F, Cariati E, Porta FA, Quici S, Roberto D (2001) Spectrochim Acta A 57:1417–1426
29. Liptay W (1974) Dipole moments and polarizabilities of molecules in excited electronic states. In: Lim EC (ed) Excited states. Academic, New York, pp 129–229
30. Bublitz GU, Boxer SG (1997) Annu Rev Phys Chem 48:213–242
31. Willetts A, Rice JE, Burland DM, Shelton DP (1992) J Chem Phys 97:7590–7599
32. Di Bella S (2002) New J Chem 26:495–497
33. Kurtz SK, Perry TJ (1968) J Appl Phys 39:3798–3813
34. Coe BJ, Chadwick G, Houbrechts S, Persoons A (1997) J Chem Soc Dalton Trans 1705–1711
35. Coe BJ, Chamberlain MC, Essex-Lopresti JP, Gaines S, Jeffery JC, Houbrechts S, Persoons A (1997) Inorg Chem 36:3284–3292
36. Coe BJ, Essex-Lopresti JP, Harris JA, Houbrechts S, Persoons A (1997) Chem Commun 1645–1646
37. Coe BJ, Harris JA, Harrington LJ, Jeffery JC, Rees LH, Houbrechts S, Persoons A (1998) Inorg Chem 37:3391–3399
38. Coe BJ, Harris JA, Asselberghs I, Persoons A, Jeffery JC, Rees LH, Gelbrich T, Hursthouse MB (1999) J Chem Soc Dalton Trans 3617–3625
39. Houbrechts S, Asselberghs I, Persoons A, Coe BJ, Harris JA, Harrington LJ, Essex-Lopresti JP (1999) Mol Cryst Liq Cryst Sci Tech B Nonlinear Opt 22:161–164
40. Houbrechts S, Asselberghs I, Persoons A, Coe BJ, Harris JA, Harrington LJ, Chamberlain MC, Essex-Lopresti JP, Gaines S (1999) Proc SPIE Int Soc Opt Eng 3796:209–218
41. Coe BJ, Harris JA, Brunschwig BS (2002) J Phys Chem A 106:897–905
42. Coe BJ, Jones LA, Harris JA, Sanderson EE, Brunschwig BS, Asselberghs I, Clays K, Persoons A (2003) Dalton Trans 2335–2341
43. Asselberghs I, Houbrechts S, Persoons A, Coe BJ, Harris JA (2001) Synth Met 124:205–207
44. Coe BJ, Houbrechts S, Asselberghs I, Persoons A (1999) Angew Chem Int Ed 38:366–369
45. Coe BJ (1999) Chem Eur J 5:2464–2471
46. Lin C-S, Wu K-C, Snijders JG, Sa R-J, Chen X-H (2002) Acta Chim Sinica 60:664–668
47. Coe BJ, Harris JA, Clays K, Persoons A, Wostyn K, Brunschwig BS (2001) Chem Commun 1548–1549
48. Coe BJ, Jones LA, Harris JA, Brunschwig BS, Asselberghs I, Clays K, Persoons A (2003) J Am Chem Soc 125:862–863
49. Coe BJ, Jones LA, Harris JA, Brunschwig BS, Asselberghs I, Clays K, Persoons A, Garín J, Orduna J (2004) J Am Chem Soc 126:3880–3891
50. Coe BJ, Harris JA, Brunschwig BS, Garín J, Orduna J, Coles SJ, Hursthouse MB (2004) J Am Chem Soc 126:10418–10427
51. Coe BJ, Jones LA, Harris JA, Asselberghs I, Wostyn K, Clays K, Persoons A, Brunschwig BS, Garín J, Orduna J (2003) Proc SPIE Int Soc Opt Eng 5212:122–136
52. Sortino S, Petralia S, Conoci S, Di Bella S (2003) J Am Chem Soc 125:1122–1123
53. Sortino S, Di Bella S, Conoci S, Petralia S, Tomasulo M, Pacsial EJ, Raymo FM (2005) Adv Mater 17:1390–1393
54. Di Bella S, Sortino S, Conoci S, Petralia S, Casilli S, Valli L (2004) Inorg Chem 43:5368–5372

55. Boubekeur-Lecaque L, Coe BJ, Clays K, Foerier S, Verbiest T, Asselberghs I (2008) J Am Chem Soc 130:3286–3287
56. Kanis DR, Lacroix PG, Ratner MA, Marks TJ (1994) J Am Chem Soc 116:10089–10102
57. Cheng LT, Tam W, Meredith GR, Marder SR (1990) Mol Cryst Liq Cryst 189:137–153
58. Cheng LT, Tam W, Eaton DF (1990) Organometallics 9:2856–2857
59. Roberto D, Ugo R, Bruni S, Cariati E, Cariati F, Fantucci PC, Invernizzi I, Quici S, Ledoux I, Zyss J (2000) Organometallics 19:1775–1788
60. Cheng LT, Tam W, Stevenson SH, Meredith GR, Rikken G, Marder SR (1991) J Phys Chem 95:10631–10643
61. Cheng LT, Tam W, Marder SR, Stiegman AE, Rikken G, Spangler CW (1991) J Phys Chem 95:10643–10652
62. Lucenti E, Cariati E, Dragonetti C, Manassero L, Tessore F (2004) Organometallics 23:687–692
63. Roberto D, Ugo R, Tessore F, Lucenti E, Quici S, Vezza S, Fantucci PC, Invernizzi I, Bruni S, Ledoux-Rak I, Zyss J (2002) Organometallics 21:161–170
64. Tessore F, Roberto D, Ugo R, Mussini P, Quici S, Ledoux-Rak I, Zyss J (2003) Angew Chem 115:472–475
65. Tessore F, Roberto D, Ugo R, Mussini P, Quici S, Ledoux-Rak I, Zyss J (2003) Angew Chem Int Ed Engl 42:456–459
66. Tessore F, Locatelli D, Righetto S, Roberto D, Ugo R, Mussini P (2005) Inorg Chem 44:2437–2442
67. Calabrese JC, Tam W (1987) Chem Phys Lett 133:244–245
68. Bourgault M, Mountassir C, Le Bozec H, Ledoux I, Pucetti G, Zyss J (1993) J Chem Soc Chem Commun 1623–1624
69. Bourgault M, Baum K, Le Bozec H, Pucetti G, Ledoux I, Zyss J (1998) New J Chem 517–522
70. Hilton A, Renouard T, Maury O, Le Bozec H, Ledoux I, Zyss J (1999) Chem Commun 2521–2522
71. Aubert V, Guerchais V, Ishow E, Hoang-Thy K, Ledoux I, Nakatani K, Le Bozec H (2008) Angew Chem Int Ed 47:577–580
72. Coe BJ, Harris JA, Jones LA, Brunschwig BS, Song K, Clays K, Garin J, Orduna J, Coles SJ, Hursthouse MB (2005) J Am Chem Soc 127:4845–4859
73. Cummings SD, Cheng LT, Eisenberg R (1997) Chem Mater 9:440–450
74. Base K, Tierney MT, Fort A, Muller J, Grinstaff MW (1999) Inorg Chem 38:287–289
75. Das S, Jana A, Ramanathan V, Chakraborty T, Ghosh S, Das PK, Bharadwaj PK (2006) J Organomet Chem 691:2512–2516
76. Maury O, Le Bozec H (2005) Acc Chem Res 38:691–704
77. Dhenaut C, Ledoux I, Samuel IDW, Zyss J, Bourgault M, Le Bozec H (1995) Nature 374:339–342
78. Le Bozec H, Renouard T, Bourgault M, Dhenaut C, Brasselet S, Ledoux I, Zyss J (2001) Synth Met 124:185–189
79. Vance FW, Hupp JT (1999) J Am Chem Soc 121:4047–4053
80. Feuvrie C, Ledoux I, Zyss J, Le Bozec H, Maury O (2005) C R Chimie 8:1243–1248
81. Le Bozec H, Renouard T (2000) Eur J Inorg Chem 229–239
82. Maury O, Viau L, Senechal K, Corre B, Guegan JP, Renouard T, Ledoux I, Zyss J, Le Bozec H (2004) Chem Eur J 10:4454–4466
83. Le Bouder T, Maury O, Le Bozec H, Ledoux I, Zyss J (2001) Chem Commun 2430–2431
84. Le Bozec H, Le Bouder T, Maury O, Bondon A, Ledoux I, Deveau S, Zyss J (2001) Adv Mater 13:1677–1681
85. Le Bozec H, Le Bouder T, Maury O, Ledoux I, Zyss J (2002) J Opt A Pure Appl Opt 4:S189–S196
86. Le Bouder T, Maury O, Bondon A, Costuas K, Amouyal E, Ledoux I, Zyss J, Le Bozec H (2003) J Am Chem Soc 125:12284–12299

87. Viau L, Bidault S, Maury O, Brasselet S, Ledoux I, Zyss J, Ishow E, Nakatany K, Le Bozec H (2004) J Am Chem Soc 126:8386–8387
88. Bidault S, Viau L, Maury O, Brasselet S, Zyss J, Ishow E, Nakatany K, Le Bozec H (2006) Adv Funct Mater 16:2252–2262
89. Coe BJ, Harris JA, Brunschwig BS, Asselberghs I, Clays K, Garn J, Orduna J (2005) J Am Chem Soc 127:13399–13410
90. Roberto D, Tessore F, Ugo R, Bruni S, Manfredi A, Quici S (2002) Chem Commun 846–847
91. Uyeda HT, Zhao Y, Wostyn K, Asselberghs I, Clays K, Persoons A, Therien MJ (2002) J Am Chem Soc 124:13806–13813
92. Tessore F, Roberto D, Ugo R, Pizzotti M, Quici S, Cavazzini M, Bruni S, De Angelis F (2005) Inorg Chem 44:8967–8978
93. De Angelis F, Fantacci S, Sgamellotti A, Cariati F, Roberto D, Tessore F, Ugo R (2006) Dalton Trans 852–859
94. Locatelli D, Quici S, Righetto S, Roberto D, Tessore F, Ashwell GJ, Amiri M (2005) Prog Solid State Chem 33:223–232
95. Sénéchal-David K, Hemeryck A, Tancrez N, Toupet L, Williams JAG, Ledoux I, Zyss J, Boucekkine A, Guégan JP, Le Bozec H, Maury O (2006) J Am Chem Soc 128:12243–12255
96. Di Bella S, Fragalà I, Ledoux I, Diaz-Garcia MA, Marks TJ (1997) J Am Chem Soc 119:9550–9557
97. Lacroix PG (2001) Eur J Inorg Chem 339–348
98. Di Bella S, Fragalà I, Ledoux I, Marks TJ (1995) J Am Chem Soc 117:9481–9485
99. Di Bella S, Fragalà I, Marks TJ, Ratner MA (1996) J Am Chem Soc 118:12747–12751
100. Di Bella S, Fragalà I, Ledoux I, Zyss J (2001) Chem Eur J 7:3738–3743
101. Di Bella S, Fragalà I (2002) New J Chem 26:285–290
102. Rigamonti L, Demartin F, Forni A, Righetto S, Pasini A (2006) Inorg Chem 45:10976–10989
103. Gradinaru J, Forni A, Druta V, Tessore F, Zecchin S, Quici S, Garbalau N (2007) Inorg Chem 46:884–895
104. Costes JP, Lamère JF, Lepetit C, Lacroix PG, Dahan F, Nakatani K (2005) Inorg Chem 44:1973–1982
105. Gaudry J-B, Capes L, Langot P, Marcén S, Kollmannsberger M, Lavastre O, Freysz E, Létard J-F, Kahn O (2000) Chem Phys Lett 324:321–329
106. Anverseng F, Lacroix PG, Malfant I, Périssé N, Lepetit C, Nakatani K (2001) Inorg Chem 40:3797–3804
107. Margeat O, Lacroix PG, Costes JP, Donnadieu B, Lepetit C, Nakatani K (2004) Inorg Chem 43:4743–4750
108. Anverseng F, Lacroix PG, Malfant I, Dahan F, Nakatani K (2000) J Mater Chem 10:1013–1018
109. Evans C, Luneau D (2002) J Chem Soc Dalton Trans 83–86
110. Rivera JM, Guzmán D, Rodriguez R, Lamère JF, Nakatani K, Santillan R, Lacroix PG, Farfán N (2006) J Organomet Chem 691:1722–1732
111. Rivera JM, Reyes H, Cortés A, Santillan R, Lacroix PG, Lepetit C, Nakatani K, Farfán N (2006) Chem Mater 18:1174–1183
112. Di Bella S, Fragalà I, Leonardi N, Sortino S (2004) Inorg Chim Acta 357:3865–3870
113. Di Bella S, Condorelli GG, Motta A, Ustione A, Cricenti A (2006) Langmuir 22:7952–7955
114. Kanis DR, Ratner MA, Marks TJ (1992) J Am Chem Soc 114:10338–10357
115. Barlow S, Bunting HE, Ringham C, Green JC, Bublitz GU, Boxer SG, Perry JW, Marder SR (1999) J Am Chem Soc 121:3715–3723
116. Barlow S, Marder SR (2000) Chem Commun 1555–1562
117. Calabrese JC, Cheng L-T, Green JC, Marder SR, Tam W (1991) J Am Chem Soc 113:7227–7232
118. Mata JA, Peris E, Asselberghs I, Van Boxel R, Persoons A (2001) New J Chem 25:299–304
119. Blanchard-Desce M, Runser C, Fort A, Barzoukas M, Lehn JM, Bloy V, Lanin V (1995) Chem Phys 199:253–261

120. Mata JA, Peris E, Asselberghs I, Van Boxel R, Persoons A (2001) New J Chem 25:1043–1046
121. Tsuboya N, Hamasaki R, Ito M, Mitsuishi M, Miyashita T, Yamamoto Y (2003) J Mater Chem 13:511–513
122. Farrell T, Meyer-Friedrichsen T, Malessa M, Haase D, Saak W, Asselberghs I, Wostyn K, Clays K, Persoons A, Heck J, Manning AR (2001) J Chem Soc Dalton Trans 29–36
123. Moore AJ, Chesney A, Bryce MR, Batsanov AS, Kelly, JF, Howard JAK, Perepichka IF, Perepichka DF, Meshulam G, Berkovic G, Kotler Z, Mazor R, Khodorkovsky V (2001) Eur J Org Chem 2671–2687
124. Janowska I, Zakrzewski J, Nakatani K, Delaire JA, Palusiak M, Walak M, Scholl H (2003) J Organomet Chem 675:35–41
125. Wrona A, Zakrzewski J, Jerzykiewicz L, Nakatani K (2008) J Organomet Chem 693:2982–2986
126. Liao Y, Eichinger BE, Firestone KA, Haller M, Luo J, Kaminsky W, Benedict JB, Reid PJ, Jen AKY, Dalton LR, Robinson BH (2005) J Am Chem Soc 127:2758–2766
127. Krishnan A, Pal SK, Nandakumar P, Samuelson AG, Das PK (2001) Chem Phys 265:313–322
128. Zhao X, Sharma HK, Cervantes-Lee F, Pannell KH, Long GJ, Shahin AM (2003) J Organomet Chem 686:235–241
129. Roy A-L, Chavarot M, Rose E, Rose-Munch F, Attias AJ, Kréher D, Fave JL, Kamierszky C (2005) C R Chimie 8:1256–1261
130. Janowska I, Zakrzewski J, Nakatani K, Palusiak M, Walak M, Sholl H (2006) J Organomet Chem 691:323–330
131. Sushanta KP, Krishnan A, Das PK, Samuelson AG (2000) J Organomet Chem 604:248–259
132. Malaun M, Kowallick R, McDonagh AM, Marcaccio M, Paul RL, Asselberghs I, Clays K, Persoons A, Bildstein B, Fiorini C, Nunzi J-M, Ward M D, McCleverty JA (2001) J Chem Soc Dalton Trans 3025–3038
133. Kumar R, Misra R, PrabhuRaja V, Chandrashekar TK (2005) Chem Eur J 11:5695–5707
134. Farrell T, Manning AR, Murphy TC, Meyer-Friedrichsen T, Heck J, Asselberghs I, Persoons A (2001) Eur J Inorg Chem 2365–2375
135. Coe BJ, Hamor TA, Jones CJ, McCleverthy JA, Bloor D, Cross GH, Axon TL (1995) J Chem Soc Dalton Trans 673–684
136. Marder SR, Perry JW, Tiemann BG, Schaefer WP (1991) Organometallics 10:1896–1901
137. Balavoine GGA, Daran J-C, Iftime G, Lacroix PG, Manoury E, Delaire JA, Maltey-Fanton I, Nakatani K, Di Bella S (1999) Organometallics 18:21–29
138. Chiffre J, Averseng F, Balavoive GGA, Daran J-C, Iftime G, Lacroix PG, Manoury E, Nakatani K (2001) Eur J Inorg Chem 2221–2226
139. Wright ME, Toplikar EG, Lackritz HS, Kerney JT (1994) Macromolecules 27:3016–3022
140. Kondo T, Horiuchi S, Yagi I, Ye S, Uosaki K (1999) J Am Chem Soc 121:391–398
141. Weidner T, Vor Der Brüggen J, Siemeling U, Träger F (2003) Appl Phys B 77:31–35
142. Malaun M, Reeves ZR, Paul RL, Jeffery JC, McCleverty JA, Ward MD, Asselberghs I, Clays K, Persoons A (2001) Chem Commun 49–50
143. Sporer C, Ratera I, Ruiz-Molina D, Zhao Y, Vidal-Gancedo J, Wurst K, Jaitner P, Clays K, Persoons A, Rovira C, Veciana J (2004) Angew Chem Int Ed 43:5266–5268
144. Cifuentes MP, Humphrey MG (2004) J Organomet Chem 689:3968–3981
145. Hurst SK, Cifuentes MP, Morrall JPL, Lucas NT, Whittall IR, Humphrey MG, Asselberghs I, Persoons A, Samoc M, Luther-Davies B, Willis AC (2001) Organometallics 20:4664–4675
146. Morrall JPL, Cifuentes MP, Humphrey MG, Kellens R, Robijns E, Asselberghs I, Clays K, Persoons A, Samoc M, Willis AC (2006) Inorg Chim Acta 359:998–1005
147. Fondum TN, Green KA, Randles MD, Cifuentes MP, Willis AC, Teshome A, Asselberghs I, Clays K, Humphrey MG (2008) J Organomet Chem 639:1605–1613
148. Cifuentes MP, Driver J, Humphrey MG, Asselberghs I, Persoons A, Samoc M, Luther-Davies B (2000) J Organomet Chem 607:72–77

149. Humphrey PA, Turner P, Masters AF, Field LD, Cifuentes MP, Humphrey MG, Asselberghs I, Persoons A, Samoc M (2005) Inorg Chim Acta 358:1663–1672
150. Cifuentes MP, Powell CE, Morral JP, McDonagh AM, Lucas NT, Humphrey MG, Samoc M, Houbrechts S, Asselberghs I, Clays K, Persoons A, Isoshima T (2006) J Am Chem Soc 128:10819–10832
151. Makowska-Janusik M, Kityk IV, Gauthier N, Frédéric P (2007) J Phy Chem C 111:12094–12099
152. Fillaut J-L, Perruchon J, Blanchard P, Roncali J, Golhen S, Allain M, Migalsaka-Zalas A, Kityk IV, Sahraoui B (2005) Organometallics 24:687–695
153. Centore R, Fort A, Panunzi B, Roviello A, Tuzi A (2004) Inorg Chim Acta 357:913–918
154. Aiello I, Caruso U, Ghedini M, Panunzi B, Quatela A, Roviello A, Sarcinelli F (2003) Polymer 44:7635–7643
155. Labat L, Lamere JF, Sasaki I, Lacroix PG, Vendier L, Asselberghs I, Perez-Moreno J, Clays K (2006) Eur J Inorg Chem 3105–3113
156. For example, see Lowry MS, Bernhard S (2006) Chem Eur J 12:7970–7977
157. De Angelis F, Fantacci S, Evans N, Klein C, Zakeeruddin SM, Moser JE, Kalyanasundaram K, Bolink HJ, Graetzel M, Nazeeruddin MK (2007) Inorg Chem 46:5989–6001
158. See for example Zhao Q, Liu S, Shi M, Wang C, Yu M, Li L, Li F, Yi T, Huang C (2006) Inorg Chem 45:6152–6160
159. Bolink HJ, Cappelli E, Coronado E, Graetzel M, Orti E, Costa RD, Viruela PM, Nazeeruddin MdK (2006) J Am Chem Soc 128:14786–14787
160. Dragonetti C, Falciola L, Mussini P, Righetto S, Roberto D, Ugo R, De Angelis F, Fantacci S, Sgamellotti A, Ramon M, Muccini M (2007) Inorg Chem 46:8533–8547
161. Dragonetti C, Righetto S, Roberto D, Ugo R, Valore A, Fantacci S, Sgamellotti A, De Angelis F (2007) Chem Commun 40:4116–4118
162. Dragonetti C, Righetto S, Roberto D, Valore A, Benincori T, Sannicolò F, De Angelis F, Fantacci S (2009) J Mater Sci Mater Electron 20:460–464
163. Calvete M, Yang GY, Hanack M (2004) Synth Met 141:231–243
164. Senge MO, Fazekas M, Notaras EGA, Blau WJ, Zawadzka M, Locos OB, Mhuircheartaigh EMN (2007) Adv Mater 19:2737–2774
165. Collini E, Mazzucato S, Zerbetto M, Ferrante C, Bozio R, Pizzotti M, Tessore F, Ugo R (2008) Chem Phys Lett 454:70–74 and references therein
166. Mc Donald IJ, Dougherty TJ (2001) J Porphyrins Phthalocyanines 5:105–129 and references therein
167. Pizzotti M, Ugo R, Annoni E, Quici S, Ledoux-Rak I, Zerbi G, Del Zoppo M, Fantucci PC, Invernizzi I (2002) Inorg Chim Acta 340:70–80
168. Annoni E, Pizzotti M, Ugo R, Quici S, Morotti T, Bruschi M, Mussini P (2005) Eur J Inorg Chem 3857–3874
169. Morotti T, Pizzotti M, Ugo R, Quici S, Bruschi M, Mussini P, Righetto S (2006) Eur J Inorg Chem 1743–1757
170. Wahab A, Bhattacharya M, Ghosh S, Samuelson AG, Das PK (2008) J Phys Chem B 112:2842–2847
171. Annoni E, Pizzotti M, Ugo R, Quici S, Morotti T, Casati N, Macchi P (2006) Inorg Chim Acta 359:3029–3041
172. Sen A, Ray PC, Das K, Krishnan V (1996) J Phys Chem 100:19611–19613
173. LeCours SM, Guan HW, DiMagno SG, Wang CH, Therien MJ (1996) J Am Chem Soc 118:1497–1503
174. Kim KS, Vance FW, Hupp JT, LeCours SM, Therien MJ (1998) J Am Chem Soc 120:2606–2611
175. Pizzotti M, Annoni E, Ugo R, Bruni S, Quici S, Fantucci PC, Bruschi M, Zerbi G, Del Zoppo M (2004) J Porphyrins Phthalocyanines 8:1311–1324
176. Yeung M, Ng ACH, Drew MGE, Vorpagel E, Breitung EM, Mc Mahon RJ, Ng DKP (1998) J Org Chem 63:7143–7150

177. De Angelis F, Fantacci S, Sgamellotti A, Pizzotti M, Tessore F, Orbelli Biroli A (2007) Chem Phys Lett 447:10–15
178. Monnereau C, Blart E, Montembault V, Fontaine L, Odobel F (2005) Tetrahedron 61:10113–10121
179. Zhang TG, Zhao Y, Asselberghs I, Persoons A, Clays K, Therien MJ (2005) J Am Chem Soc 127:9710–9720
180. Zhang TG, Zhao Y, Song K, Asselberghs I, Persoons A, Clays K, Therien MJ (2006) Inorg Chem 45:9703–9712
181. Bonifassi P, Ray PC, Leszczynski J (2006) Chem Phys Lett 431:321–325
182. Ray PC, Bonifassi P, Leszczynski J (2008) J Phys Chem A 112:2870–2879
183. Liao MS, Bonifassi P, Leszczynski J, Huang MJ (2008) Mol Phys 106:147–160
184. Duncan TV, Song K, Hung ST, Miloradovic I, Nayak A, Persoons A, Verbiest T, Therien MJ, Clays K (2008) Angew Chem Int Ed 47:2978–2981
185. De la Torre G, Vásquez P, Agulló-López F, Torres T (2004) Chem Rev 104:3723–3750
186. Chen Y, Hanack M, Blau WJ, Dini D, Liu Y, Lin Y, Bai J (2006) J Mater Sci 41:2169–2185
187. Martín G, Martínez-Díaz MV, De la Torre G, Ledoux I, Zyss J, Agulló-López F, Torres T (2003) Synth Met 139:95–98
188. Maya EM, García-Frutos EM, Vásquez P, Torres T, Martín G, Rojo G, Agulló-López F, González-Jonte RH, Ferro VR, García de la Vega JM, Ledoux I, Zyss J (2003) J Phys Chem A 107:2110–2117
189. De la Escosura A, Claessens CG, Ledoux-Rak I, Zyss J, Martínez-Díaz MV, Torres T (2005) J Porphyrins Phthalocyanines 9:788–793
190. Claessens CG, González-Rodríguez D, Torres T, Martín G, Agulló-López F, Ledoux I, Zyss J, Ferro VR, García de la Vega JM (2005) J Phys Chem B 109:3800–3806
191. Li CQ, Manaka T, Iwamoto M (2003) Thin Solid Films 438/439:162–166
192. Cheng X, Yao S, Li C, Manaka T, Iwamoto M (2003) Sci Chin 46:379–386
193. Li CQ, Manaka T, Iwamoto M (2004) Jpn J Appl Phys 43:2330–2334
194. Kanis DR, Ratner MA, Marks TJ (1992) J Am Chem Soc 114:10338–10357
195. Behrens U, Brussaard H, Hagenau U, Heck J, Hendrickx E, Kornich J, van der Linden JGM, Persoons A, Spek AL, Veldman N, Voss B, Wong H (1996) Chem Eur J 2:98–103
196. Mata J, Uriel S, Peris E, Llusar R, Houbrechts S, Persoons A (1998) J Organomet Chem 562:197–202
197. Briel O, Sünkel K, Krossing I, Nöth H, Schmälzlin E, Meerholz K, Brächle C, Beck W (1999) Eur J Inorg Chem 483–490
198. Jayaprakash KN, Ray PC, Matsuoka I, Bhadbhade MM, Puranik VG, Das PK, Nishihara H, Sarkar A (1999) Organometallics 18:3851–3858
199. Houbrechts S, Clays K, Persoons A, Cadierno V, Pilar Gamasa M, Gimeno J (1996) Organometallics 15:5266–5268
200. Vance FW, Karki L, Reigle JK, Hupp JT, Ratner MA (1998) J Phys Chem A 102:8320–8324
201. Pizzotti M, Ugo R, Roberto D, Bruni S, Fantucci PC, Rovizzi C (2002) Organometallics 21:5830–5840
202. Bruschi M, Fantucci PC, Pizzotti M (2005) J Phys Chem A 109:9637–9645
203. Pizzotti M, Ugo R, Dragonetti C, Annoni E, Demartin F, Mussini P (2003) Organometallics 22:4001–4011
204. Coe BJ, Fitzgerald EC, Helliwell M, Brunschwig BS, Fitch AG, Harris JA, Coles SJ, Horton PN, Hursthouse MB (2008) Organometallics 27:2730–2742
205. Singer KD, Sohn JE, King LA, Gordon HM, Katz HE, Dirk CW (1989) J Opt Soc Am B 6:1339–1350
206. Dirk CW, Katz HE, Schilling ML, King LA (1990) Chem Mater 2:700–705

NLO Molecules and Materials Based on Organometallics: Cubic NLO Properties

Mark G. Humphrey, Marie P. Cifuentes, and Marek Samoc

Abstract Relevant background theory, experimental procedures, and significant recent advances in the third-order NLO properties of organometallic complexes are reviewed, with particular emphasis on spectral dependence studies and switching of nonlinearity.

Keywords Cubic nonlinearity, Molecular switches, Nonlinear absorption, Nonlinear refraction

Contents

1 Introduction 58
2 Theory and Experiment in Third-Order NLO of Organometallics 58
   2.1 Theory of Third-Order NLO Effects 58
   2.2 Experiment in Third-Order NLO Studies 62
3 Structure-Property Developments Since 2000 65
4 Spectral Dependencies 67
5 Switching 69
6 Materials 71
7 Conclusion 71
References 72

M.G. Humphrey (✉) and M.P. Cifuentes
Research School of Chemistry, Australian National University, Canberra, ACT 0200, Australia
e-mail: Mark.Humphrey@anu.edu.au, Marie.Cifuentes@anu.edu.au

M. Samoc
Institute of Physical and Theoretical Chemistry, Wroclaw University of Technology, 50–370 Wroclaw, Poland
e-mail: Marek.Samoc@pwr.wroc.pl

1 Introduction

Nonlinear optical (NLO) effects originate from high-intensity electric fields (such as those from laser beams) interacting with matter. These interactions can result in the electromagnetic field components of the incident light beam being modified, and new components (differing in phase, frequency, etc.) being generated. These effects can have a range of applications in laser technologies, optical signal processing, multiphoton information storage and retrieval, microscopy, and multiphoton photodynamic therapy, for example.

It can be challenging to meet the demands on prospective materials for device applications. Apart from the obvious requirement that NLO responses be sufficient for the specific application, the material needs to tolerate the device manufacturing and operating conditions (e.g., it needs to possess a sufficiently high damage threshold), as well as having acceptable photochemical and thermal stability or fluorescence efficiency (depending on the application). While crystals of inorganic salts and glasses currently dominate the market for second- and third-order NLO applications, respectively, there are shortcomings with aspects of their performance that have focused attention on organics and, more recently, organometallics as potential NLO materials. In particular, organometallics have the potential to combine the advantages of organics (such as fast response, ease of processing, and considerable design flexibility) with the benefits that may flow from incorporation of a metal center (such as variable oxidation state and coordination geometry).

In contrast to second-order nonlinearities, for which considerable success defining structure-property relationships has been achieved, considerably less is known of how to optimize third-order effects. Third-order optical nonlinearities of organometallic complexes were first reported in the mid-1980s. The promising early studies stimulated considerable interest, and an ever-increasing number of reports, the results from which have been summarized in several reviews that also include introductions to the background theory of nonlinear optics [1–8]. This chapter includes a brief overview of theory relevant to cubic NLO and a summary of experimental procedures to measure cubic nonlinearities, followed by a brief review highlighting the significant advances in the field over the past few years; while theoretical studies of the NLO properties of organometallics have attracted interest [9, 10], the emphasis here is on experimental outcomes.

2 Theory and Experiment in Third-Order NLO of Organometallics

2.1 Theory of Third-Order NLO Effects

When a molecule is exposed to an electromagnetic field, it is usually the electric component of the field that is of much greater significance than the magnetic component, and the response of the molecule is usually treated within a dipolar approximation in which the whole interaction is described as the time evolution of a point dipole. Both of these approximations have limited value; for example, neglecting the magnetic interactions does not allow for the description of chirooptic effects, and assuming a point character of the dipole and neglecting the multipolar character of the effects may be inappropriate for large molecules such as dendrimers. However, the traditional approach to explaining the origin of NLO effects on the molecular scale is by simply assuming that an electric field of magnitude E creates a point dipole μ in a molecule, and this dipole can be expanded in series against the powers of E. Because μ and E are vectors, the equation relating them must take into account dependences of all of the Cartesian components of both vectors. Using Einstein's convention of summing over repeated indices and employing the esu system of units, one can write[1]:

$$\mu_i = \mu_i(0) + \alpha_{ij}E_j + \beta_{ijk}E_jE_k + \gamma_{ijkl}E_jE_kE_l + \ldots, \tag{1}$$

where μ_i are components of the total dipole moment, $\mu_i(0)$ is the dipole moment at low field strengths (the permanent dipole moment), α_{ij} is the molecular polarizability, β_{ijk} is the first hyperpolarizability (also called the second-order or quadratic polarizability), and γ_{ijkl} is the second hyperpolarizability (also called the third-order or cubic polarizability)[2]. The order of expansion in Eq. (1) determines the naming of the effects that are due to the consecutive terms: linear optical effects are those that are described by the linear polarizability α, quadratic NLO effects are due to β, and cubic NLO effects to γ.

Strictly speaking, Eq. (1) is only valid for static electric fields. For any field that is time dependent, i.e., $E(t)$, the molecular response $\mu(t)$ will always lag behind somewhat. This is taken into account by assuming that α, β, and γ are complex quantities, and that they relate Fourier components of the time-varying dipole moment to Fourier components of the time-varying electric field. In effect, Eq. (1) needs to be rewritten for the combination of field amplitudes and field frequencies that are of interest. Limiting the following to cubic NLO effects, such an equation will have the following form:

$$\mu_i^{(3)}(\omega_4) = \gamma_{ijkl}(\omega_4;\omega_1,\omega_2,\omega_3)E_j(\omega_1)E_k(\omega_2)E_l(\omega_3), \tag{2}$$

where an interaction of three fields at three frequencies ω_1, ω_2, and ω_3 leads to an induced nonlinear component of a dipole oscillating at $\omega_4 = \omega_1 + \omega_2 + \omega_3$.

---

[1] Note that the equations describing nonlinear optical effects can have different forms depending on the system of units used (esu or SI), the inclusion of the $1/n!$ factors of the power expansion in the coefficients or using them in front of the coefficients, and inclusion (or not) of the degeneracy factors that may appear in the equations, depending on the number of identical fields in a given nonlinear interaction. See, for example, [11].

[2] Note that some authors refer to β as the *first-order hyperpolarizability* and γ as the *second-order hyperpolarizability*. While this is formally correct, in the opinion of the present authors such nomenclature should be avoided since it can lead to confusion as to the order of the nonlinear process that is being described.

It should be noted that γ is a tensor with four indices, and thus has 81 components; the values of these components can depend on any of the frequencies in Eq. (2). Fortunately, symmetry considerations can reduce the number of independent tensor components considerably. In addition, the dispersion of γ can sometimes be simplified when specific NLO interactions are considered.

Cubic NLO processes can be termed *four-wave mixing* processes, because in each case described by Eq. (2), one deals with three input fields and an output field (oscillating at ω_4), the latter being generated by the nonlinear component of the induced dipole moment. Among the examples of wave mixing that are of practical importance, the simplest one occurs when only a single frequency field originating from a single light beam is present. In this case, the field is described by:

$$E(t) = \frac{1}{2}E(\omega)[\exp(-i\omega t) + \exp(i\omega t)] \tag{3}$$

and the cubic nonlinear interaction has only two forms: either the frequencies of the three amplitudes in Eq. (2) sum to form the third-harmonic of the fundamental frequency, i.e., $\omega_4 = 3\omega = \omega + \omega + \omega$, or the interaction involves one of the amplitudes taken with the reverse phase (or, formally, as a negative frequency $-\omega$), so $\omega_4 = \omega = \omega - \omega + \omega$. The former case is called third-harmonic generation (THG). The latter case is sometimes called self-action, and only introduces changes to the amplitude and phase of the existing electromagnetic field at ω. These changes are nevertheless of great practical significance, since they lead to effects such as self-focusing, soliton formation, all-optical switching, and nonlinear absorption processes such as two-photon absorption.

To extend the discussion of cubic optical nonlinearity from the molecular scale to the macroscopic (bulk) scale, one needs to consider the additivity of the individual molecular tensors and their respective orientation in space, as well as local field corrections. Dealing with hyperpolarizability tensors is greatly simplified when the nonlinear material under consideration is a collection of randomly oriented molecules, such as those in a liquid solution, or those in a solid solution, e.g., a molecularly-doped polymer, a polymer with nonlinear chromophores present as side groups, a sol–gel glass or ormosil containing chromophores, etc. Averaging of the γ_{ijkl} tensor over all orientations leads to an average $\langle \gamma \rangle$; for the case of a single electric field component acting on an assembly of nonlinear chromophores, $\langle \gamma \rangle$ is given by [12]:

$$\langle \gamma \rangle = \frac{1}{15}\gamma_{ijkl}(\delta_{ij}\delta_{kl} + \delta_{ik}\delta_{jl} + \delta_{il}\delta_{jk}), \tag{4}$$

where δ_{ij} is the substitution tensor (unity if $i = j$, zero otherwise). When the γ tensor is dominated by a single component γ_{1111}, this expression gives $\langle \gamma \rangle = 1/5\gamma_{1111}$. It should be noted that it is the average value of γ that is normally quoted as the measurement result when experiments are carried out on solutions of chromophores. The appropriate macroscopic measure of the cubic optical nonlinearity is the cubic nonlinear susceptibility $\chi^{(3)}(\omega_4;\omega_1,\omega_2,\omega_3)$, which can be defined by macroscopic equations analogous to Eqs. (1) and (2) by considering the macroscopic analogue of the dipole moment, the polarization vector P, and its appropriate Fourier components. For a solution, the relation between $\chi^{(3)}$ and $\langle\gamma\rangle$ can be written as:

$$\chi^{(3)} = L^4 \sum_r N_r \langle\gamma\rangle_r, \tag{5}$$

where the index r extends over all components of the solution (the solvent as well as the solutes), N_r represents the concentration of a component molecule (in molecules cm^{-3}), and L is the local field correction factor, usually approximated by a Lorentz expression $L = (n^2+2)/3$, n being the refractive index. Similar to γ, $\chi^{(3)}$ is generally complex, i.e., it has real and imaginary parts:

$$\chi^{(3)} = \text{Re}(\chi^{(3)}) + i\text{Im}(\chi^{(3)}) = \chi^{(3)}_{\text{real}} + i\chi^{(3)}_{\text{imag}}. \tag{6}$$

The third-harmonic generation (THG) process that is due to the cubic susceptibility $\chi^{(3)}(3\omega;\omega,\omega,\omega)$ is a general effect, i.e., it is exhibited by all matter, the contribution to THG from air being of particular importance in some experiments. Although THG may lead to up-conversion of laser beam frequency (e.g., from infrared at 1,064 nm to ultraviolet at 355 nm), the practical use of the process is difficult because of the near impossibility to equalize the velocity of the fundamental and third-harmonic waves in typical NLO materials. In practice, the third-harmonic of laser beams is obtained by a different route: doubling of frequency ω and then frequency summation of 2ω and ω in two subsequent quadratic NLO processes. However, THG is of importance, not only for testing of NLO properties of molecules, but also for laser diagnostics and for microscopic imaging of various objects including biological ones.

In contrast, the degenerate cubic NLO interaction, i.e., one in which the result of the interaction of three field components at ω is also at the same frequency ω, has numerous important applications. The notion that this interaction modifies amplitude and/or phase of the existing wave may be rephrased taking into account that the change in amplitude occurring at a certain distance is equivalent to absorption of light while the change in phase means slowing down or speeding up the wave, or, indeed, modification of the effective refractive index of the medium. Therefore, two practical effects of the degenerate cubic NLO interaction are described in terms of two simple quantities: the nonlinear refractive index n_2, defined by:

$$n(I) = n_0 + n_2 I \tag{7}$$

and the nonlinear absorption coefficient α_2, defined by:

$$\alpha(I) = \alpha_0 + \alpha_2 I, \tag{8}$$

where I represents the light intensity (related to the square of the electric field amplitude), and n_0 and α_0 are the refractive index and the absorption coefficient at low light intensities, respectively. It can be shown that the nonlinear refractive index is related to the real part of $\chi^{(3)}$ by [13]:

$$n_2 = \frac{4\pi}{n_0 c} n_2' = \frac{12\pi^2}{n_0^2 c} \text{Re}\left[\chi_{xxxx}^{(3)}(\omega;\omega,-\omega,\omega)\right]. \tag{9}$$

[Note that n_2' is a differently defined nonlinear index that relates the refractive index change to the square of the electric field amplitude]. The nonlinear absorption coefficient is related to the imaginary part of $\chi^{(3)}$ through [13]:

$$\alpha_2 = \frac{48\pi^3}{n_0^2 c \lambda} \text{Im}\left(\chi_{xxxx}^{(3)}(\omega;\omega,-\omega,\omega)\right). \tag{10}$$

It is often convenient to treat the refractive index of a medium as a complex quantity $\hat{n} = n + ik$, the real part of it being responsible for refraction and the imaginary part being responsible for absorption. Extending this to nonlinear phenomena, one can also treat the nonlinear index as complex and describe nonlinear refraction as being due to the real part of n_2 and nonlinear absorption as due to the imaginary part. An alternate way of expressing the nonlinear absorption properties of a molecule is by defining its nonlinear absorption cross-section. The two-photon absorption cross-section σ_2 is related to the imaginary part of γ of a molecule, and can be calculated from the nonlinear absorption coefficient α_2:

$$\sigma_2 = \frac{\hbar\omega}{N}\alpha_2, \tag{11}$$

where N is the concentration of the absorbing molecules.

2.2 Experiment in Third-Order NLO Studies

Determination of cubic NLO properties of molecules is most often carried out in liquid solutions, but it is possible to use solid samples made by (for example) dispersing the chromophore in a host polymer, or as thin films of the pure compound under investigation, if the optical quality of the samples is sufficiently high. Depending on the method of measurement, the real and imaginary parts of $\chi^{(3)}$ (or n_2 and α_2) may be determined, but in some cases only the modulus $|\chi^{(3)}|$ is measured. The experimental techniques have been described in detail elsewhere [13], so only a short description of them is given here.

2.2.1 Third-Harmonic Generation

THG experiments remain popular for the estimation of cubic molecular nonlinearities. The experiment is usually carried out by recording the intensity of the third-harmonic generated by a beam from an infrared laser. Because of the many uncertainties related to the details of the generation of the third-harmonic, an often-adopted version of the experiment is carried out by recording fringes that appear when a thin film sample deposited on a glass plate is rotated out of the plane perpendicular to the laser beam. The resulting changes in the optical paths for the fundamental and third-harmonic beams lead to constructive or destructive interference. The fringe pattern thus obtained can be compared with one obtained for the glass plate alone, to evaluate both the amplitude and phase of $\chi^{(3)}(3\omega;\omega,\omega,\omega)$ for the thin film sample, which lead to the real and imaginary parts of $\chi^{(3)}$. A serious limitation of the THG technique when applied to organometallics is that $\chi^{(3)}(3\omega;\omega,\omega,\omega)$ may be quite different from $\chi^{(3)}(\omega;\omega,-\omega,\omega)$, which as discussed above has more practical significance – this is due to the fact that different resonances of the molecule contribute to these two susceptibilities (resonances close to ω and 3ω in the case of THG and resonances near ω and 2ω in the case of the degenerate susceptibility).

2.2.2 Z-Scan

Z-scan [14] (Fig. 1) is by far the most popular technique for investigations of the cubic molecular nonlinearities of organometallics, and is typically undertaken

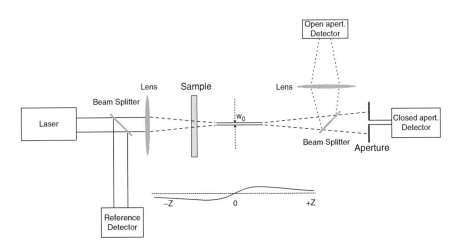

Fig. 1 Scheme of the Z-scan experiment in which the closed-aperture scan and the open-aperture scan are carried out simultaneously

with the organometallics in solutions of organic solvents. This ingenious technique consists in focusing a laser beam with a lens, scanning the sample to be investigated along the beam axis (the z-axis, hence the name Z-scan) in the range from a few Rayleigh lengths before the focal plane to a few Rayleigh lengths after the focal plane, and observing changes in the transmitted beam in the far field as a function of the position of the sample, z (the Rayleigh length is given by $z_R = \pi w_0^2/\lambda$, where w_0 is the laser beam spot size at focus). A very simple way of detecting distortion of the beam due to the self-focusing/self-defocusing effects imposed on it by the sample is by placing a small aperture in the far field and recording the power transmitted through this aperture as a function of z. The sample starts its travel at the point where the beam is still relatively large and therefore the light intensity is low, but as it comes closer to the focal plane at $z = 0$, the intensity increases and is then reduced again. These intensity variations lead to formation in the sample of an induced positive (self-focusing, $n_2 > 0$) or negative (self-defocusing, $n_2 < 0$) lens whose effect is to add to that of the primary lens and to modify the shape of the beam in the far field. As the beam becomes larger or smaller through focusing/defocusing, the aperture transmittance increases or decreases, and these variations (in the so-called closed-aperture scan) can be compared to those predicted from a theoretical computation.

While self-focusing and self-defocusing are manifestations of the refractive part of the degenerate cubic optical nonlinearity, the absorptive part results in variation of the total power of the beam transmitted through the sample as a function of z. This can be monitored with a detector that integrates the power in the whole beam, and the changes of such power as a function of z (in the so-called open-aperture scan) can be directly related to the nonlinear absorption coefficient of the sample, α_2.

2.2.3 Nonlinear Absorption and Nonlinear Absorption-Induced Fluorescence

Open-aperture Z-scan is often used to measure nonlinear absorption alone, particularly in cases where such absorption is more complicated than a simple two-photon process. Alternatively, one can simply measure the transmittance of a sample as a function of the incident laser power (by attenuation of the laser beam, without changing its focusing), and then deduce the nonlinear absorption parameters from the deviation from linearity of the transmission vs power plot. A very convenient method of measuring two-photon absorption cross-section is where the degree of nonlinear absorption is judged from the amount of two-photon-induced fluorescence emitted by a sample exposed to short laser pulses in a wavelength range where there is no one-photon absorption. This popular technique is, however, seldom applicable to organometallics, for which fluorescence is commonly strongly quenched by the metal atoms.

2.2.4 Degenerate Four-Wave Mixing and Pump-Probe Measurements

One of the crucial problems in the determination of cubic optical nonlinearity parameters is whether the effects measured are really due to the electronic hyperpolarizabilities of molecules, or if they are a result of more complicated NLO processes which may involve such effects as multistep multiphoton absorption (i.e., one-photon absorption followed by excited-state absorption), nonlinear refraction due to the presence of short or long-lived excitation, or even thermal effects due to the deposition of heat by linear and nonlinear absorption processes occurring in the sample under investigation. The Z-scan technique is not suitable for resolving issues of this kind (although it is often assumed that the use of low repetition rate femtosecond laser pulses minimizes the contribution of the unwanted cumulative effects). Time-resolved techniques can be employed instead. The principle of these techniques is that a beam (or a pair of beams) carrying short (femtosecond or picosecond) laser pulses is directed onto a sample and produces a transient change in the optical properties of the sample, i.e., it changes either its refractive (refractive index) or absorptive properties. In the case of degenerate four-wave mixing (DFWM), the two beams used as the pump interfere in the sample, and actually create a volume grating of the absorptive and/or refractive properties modification. A probe beam is directed at the sample after a delay, the delay being produced by passing the probe beam through a slightly longer path than that of the pump beam(s). The effect on the probe beam is that its transmission may vary depending on the degree of the change induced in the sample, or it may be diffracted to a certain degree, both depending on the strength of the existing induced grating in the sample. The measurements of these changes in the probe beam or the diffracted beam, as a function of the delay of the pump compared to the probe, allow one to estimate the magnitude of the nonlinear effects and to evaluate their temporal evolution, and thereby allow one to conclude whether the effects are essentially instantaneous or if they are due to processes having a certain lifetime.

3 Structure-Property Developments Since 2000

There has been considerable activity in correlating structural changes to the magnitude of optical nonlinearities for various types of organometallics in the past few years, but ferrocene complexes and metal alkynyl complexes continue to be the most intensively studied classes of organometallics in NLO. While earlier studies focused on simple compounds coupling ferrocenyl groups to organic π-delocalizable groups, ferrocenyl units have now been used to construct organometallic-coordination hybrid complexes such as **1** [15]. Z-scan studies at 532 nm using 8 ns pulses for **1**, the ferrocenyl ligand, and related mercury- and cadmium-containing derivatives suggest that the nonlinearity derives from the ferrocenyl ligand and that there is little contribution from the posttransition metal. The reported γ values are five to six orders of magnitude larger than previously reported data for ferrocenyl complexes obtained with femtosecond pulses, so there are likely to be contributions from thermal effects, photochemical changes, and other cumulative effects, particularly excited-state absorption.

Most of the earlier studies of ruthenium alkynyl complexes bearing acceptor substituents involved 4-nitrophenyl groups on the alkynyl ligand. This has now been extended to complexes **2–4** bearing barbiturate acceptor groups, which were assessed by DFWM at 532 nm using picosecond pulses [16]. As is the case with 4-nitro-containing complexes, nonlinearity increases on π-bridge lengthening, with complex **4** the most active of the complexes examined.

There has been enormous interest in the optical limiting properties of transition metal clusters in the past decade. Most data result from Z-scan studies with ns pulses at 532 nm. For example, the half-open cubane-like clusters $WCu_3(\mu_3\text{-}S)_3(\mu\text{-}Br)Br(EPh_3)_2(\eta^5\text{-}C_5Me_5)$ (E = P, As (**5**)) show similar threshold limiting values, with the phosphine-containing cluster revealing somewhat greater nonlinear absorption at the same concentration [17]. The cyclic complex $(\eta^5\text{-}C_5Me_5)(S)W\{(\mu\text{-}S)Au(\mu\text{-}S)_2W(\eta^5\text{-}C_5Me_5)(\mu\text{-}S)\}Au(\mu\text{-}S)$ (**6**) in which the metal atoms are held together solely by bridging sulfido ligands was examined by a combination of ps DFWM and ns Z-scan studies [18]. While the reported nonlinearities are large, these data were collected at 532 nm, a wavelength that corresponds to moderate linear absorption for this compound and indeed most of the clusters to have been examined thus far. Studies of clusters over a broad wavelength range are lacking. Since the optical limiting phenomenon occurring in the nanosecond domain is most often due to a combination of processes, involving among others one-photon absorption, two-photon absorption, and excited state absorption, the lack of wide wavelength and time-resolved studies restricts the possibilities of any generalization of the results or, in fact, correlation between the structural features of the investigated molecules, their photophysical properties, and the power limiting merit.

4 Spectral Dependencies

The vast majority of reports of the third-order NLO properties of organometallics have focused on results from a single laser irradiation wavelength. These data are most useful when comparing efficiency for specific applications at that wavelength. The problem with this approach is that contrasting data obtained for a range of complexes at a single wavelength rarely affords a structure-property outcome similar to that obtained from a comparison of the maximal values from a wide-spectral range study. Difficulty in access to wide-spectral range NLO data has retarded the development of this area. Almost all of the spectral dependencies as currently exist for organometallics have been obtained from a tedious point-by-point data collection using the Z-scan technique and a tunable light source. The recently-developed techniques of white-light continuum Z-scan and white-light continuum pump-probe have been applied to organic molecules [19–24], inorganic complexes [25–27], inorganic semiconductors [28, 29], and organic polymers [30, 31], but the only application to organometallics thus far is to (η^5-cyclopentadienyl)(η^6-cumene)iron (III) hexafluorophosphate (Irgacure 261), a commercially-available photo-initiator, whose maximal 2PA cross-section was found to be very low and could only be defined in terms of an upper bound (<20 GM) [20]. It can confidently be anticipated that application of this new technique to organometallics will expand in the future.

Thus far, there has only been one study of a systematically-varied series of organometallic complexes. Femtosecond Z-scan measurements of the platinum-terminated polyynes trans,trans-{(p-MeC$_6$H$_4$)$_3$P}$_2$(p-MeC$_6$H$_4$)Pt(C≡C)$_n$Pt(p-C$_6$H$_4$Me){P(p-C$_6$H$_4$Me)$_3$}$_2$ (7) (n = 3–6, 8, 10, 12) afforded two-photon absorption maxima that were shown to red-shift upon chain lengthening [32].

Extrapolation of the sp-carbon chain length dependence of nonlinear absorption maxima permitted an estimate (neglecting saturation) of 1,000 nm for that of the infinite carbon chain, carbyne. The maximal values increase superlinearly on chain lengthening, and can be fitted to a power law with an exponent of ca. 1.8, suggesting that longer chain examples than are currently experimentally accessible will have extraordinarily large σ_2 values.

Two similar alkynylruthenium dendrimers with nitro groups at the periphery have been shown to have very different nonlinear absorption behavior. Dendrimer **8**, which possesses ligated ruthenium units at the outermost branches only, revealed dispersion of the cubic nonlinearity that could be modeled by simple relations assuming competition between two-photon absorption and absorption saturation [33]. Dendrimer **9** has ligated ruthenium groups at both the core and periphery of the dendrimer. While the short wavelength behavior (625–950 nm) corresponds to a two-photon absorption process, the longer wavelength profile (1,000–1,300 nm) is consistent with the dominance of three-photon absorption, and the dendrimer possesses a record 3PA coefficient [34]. The advantages that accrue from exploiting

9

multiphoton processes (superior resolution for applications requiring spatial control, and the opportunity to use longer and therefore technologically-desirable wavelengths) are accentuated when proceeding from two-photon absorption to three-photon absorption, but little is known of how to optimize the 3PA coefficient. This field is still in its infancy – it is to be hoped that further reports of such fifth-order (quintic) nonlinearity are forthcoming for organometallics in the near future.

5 Switching

The magnitudes of optical nonlinearities of organometallics approach those of the best organics, but to justify the possible additional costs associated with organometallics (such as potentially expensive metals and ligands, less stable compounds, lower-yielding and longer synthetic procedures) necessitates "value-adding" to what is possible with purely organic compounds. One area in which organometallics may be superior to organics is modulating or switching nonlinearity. The most popular approaches to reversible switching of molecular NLO properties are by protonation/deprotonation, oxidation/reduction, or photoisomerization procedures [35]. Of these three approaches, the ready accessibility of metals in multiple stable oxidation states suggests that redox switching of optical nonlinearity may be an area in which organometallics are preferred. The cubic nonlinearities of several ruthenium alkynyl complexes and dendrimers have been switched in situ in a modified optically-transparent thin-layer electrochemical (OTTLE) cell [36–39], with several examples corresponding to reversing the sign and magnitude of nonlinear refraction and nonlinear absorption, and a specific example corresponding to a cubic NLO "on/off" switch. Incorporation of metals with widely differing oxidation potentials into the one NLO-active molecule affords the prospect of multiple NLO states. Applying an appropriate potential to the heterobimetallic complex **10** allows one to access the Fe^{II}/Ru^{II}, Fe^{III}/Ru^{II}, and Fe^{III}/Ru^{III} states, all of which have distinct linear optical and NLO properties; at 790 nm, open-aperture Z-scan studies reveal that the Fe^{II}/Ru^{II}, Fe^{III}/Ru^{II}, and Fe^{III}/Ru^{III} states correspond to effectively zero nonlinear absorption, two-photon absorption, and saturable absorption, respectively, or zero, positive, and negative values of the imaginary component of the cubic nonlinearity [40].

Switching the cubic nonlinearity of ruthenium alkynyl complexes by a protonation/deprotonation sequence (via a vinylidene complex) was demonstrated by fs Z-scan studies at 800 nm several years ago [41]. Recently, protic and electrochemical switching were demonstrated in the ruthenium alkynyl cruciform complex **11** for which distinct linear optical and NLO behavior were noted for the vinylidene complex and the Ru(II) and Ru(III) alkynyl complexes [42]. Because the oxidation/reduction and protonation/deprotonation procedures are independent, this system corresponds to switching by "orthogonal" stimuli.

6 Materials

The major focus of third-order NLO studies with organometallics over the past few years has continued to be solution measurements, which permit assessment of molecular nonlinearities and development of structure-property relationships; reports of bulk susceptibilities of organometallic-containing films are comparatively rarer. Bis(arene)chromium complexes have been incorporated into cyano-containing polymeric matrices and very high purely electronic third-order NLO susceptibilities of the resultant polymeric films have been noted from Z-scan and spectrally-resolved two-beam coupling studies [43]. The molar ratio of Cr:monomer was as high as 1:4, and the resultant homogeneous oligomeric product, which is produced by cyanoethylation of the metal-bound arene rings, was assessed as containing ca. eight monomer units. The Z-scan studies (utilizing 40 ps pulses at 1,064 nm or 3 ps pulses at 1,054 nm) afforded $\chi^{(3)}$ values as high as -1.7×10^{-10} esu, of the same order as the best organic polymer materials. Spectrally-resolved two-beam coupling using fs pulses and a central wavelength of 795–800 nm confirmed that the test composites exhibit significant ultrafast electronic nonlinearity.

The alkynylruthenium complexes **12–14** were spin-coated from dichloromethane solution and the resultant ca. 0.2 μm thick films assessed by THG using 15 ps pulses at 1,064 nm, with a ca. 50% increase in $\chi^{(3)}$ value noted on proceeding from **12** to **13** and **14** consistent with an extended π-system for the latter complexes involving the *trans*-disposed alkynyl ligands [44].

7 Conclusion

In contrast to the rapid development of structure–property relationships for quadratic optical nonlinearities of organometallics, the effect of molecular variation on cubic NLO properties has been slow to be systematized, a shortcoming that access to broad-wavelength-range fs sources should progressively address. While initial emphasis with organometallics was in using the metal center to stabilize reactive organics and afford unusual geometries and thereby charge distributions, recent studies have focused on exploiting access to multiple stable oxidation states, thereby affording access to new types of molecular NLO switches.

Materials development exploiting cubic NLO properties of organometallics has been slow, reflecting the fact that the focus is still on molecular rather than bulk material properties. It is to be hoped, however, that future studies will involve the organometallics with the most promising molecular NLO properties and will attempt to use these to meet the NLO material challenges of nanophotonics and biophotonics, e.g., the need for highly efficient nonlinear absorbers for use in nanofabrication, 3D memories, and bioimaging.

References

1. Marder SR, Sohn JE, Stucky GD (1991) Materials for nonlinear optics, chemical perspectives. ACS, Washington DC
2. Marder SR (1992) In: Bruce DW, O'Hare D (eds) Inorganic materials. Wiley, Chichester, UK
3. Long NJ (1995) Angew Chem Int Ed Engl 34:21–38
4. Whittall IR, McDonagh AM, Humphrey MG, Samoc M (1999) Adv Organomet Chem 43: 349–405
5. Gray GM, Lawson CM (1999) In: Roundhill DM, Fackler JP Jr (eds) Optoelectronic properties of inorganic compounds. Plenum, New York, NY
6. Kershaw SV (1999) In: Roundhill DM, Fackler JP Jr (eds) Optoelectronic properties of inorganic compounds. Plenum, New York, NY
7. Long NJ (1999) In: Roundhill DM, Fackler JP Jr (eds) Optoelectronic properties of inorganic compounds. Plenum, New York, NY
8. Morrall JP, Dalton GT, Humphrey MG, Samoc M (2008) Adv Organomet Chem 55:61–136
9. Roy S, Kulshrestha K (2005) Opt Commun 252:275–285
10. Liu Y-C, Kan Y-H, Wu S-X, Yang G-C, Zhao L, Zhang M, Guan W, Su Z-M (2008) J Phys Chem A 112:8086–8092
11. Shi RF, Garito AF (1998) In: Kuzyk MG, Dirk AF (eds) Characterization techniques and tabulations for organic nonlinear optical materials. CRC, Boca Raton, USA
12. Buckingham AD, Pople JA (1955) Proc Phys Soc A 68:905–909
13. Sutherland RL (2003) Handbook of nonlinear optics, 2nd edn. Marcel Dekker, New York
14. Sheikh-bahae M, Said AA, Wei T, Hagan DJ, Van Stryland EW (1990) IEEE J Quantum Electron 26:760–769
15. Li G, Song Y, Hou H, Li L, Fan Y, Zhu Y, Meng X, Mi L (2003) Inorg Chem 42:913–920
16. Luc J, Migalska-Zalas A, Tkaczyk S, Andriès J, Fillaut J-L, Meghea A, Sahraoui B (2008) J Optoelectron Adv Mater 10:29–43
17. Lang J-P, Sun Z-R, Xu Q-F, Yu H, Tatsumi K (2003) Mater Chem Phys 82:493–498
18. Lang J-P, Yu H, Ji S-J, Sun Z-R (2003) Phys Chem Chem Phys 5:5127–5132
19. De Boni L, Andrade AA, Misoguti L, Mendonça CR, Zilio SC (2004) Opt Express 12: 3921–3927
20. Schafer KJ, Hales JM, Balu M, Belfield KD, Van Stryland EW, Hagan DJ (2004) J Photochem Photobiol A Chem 162:497–502
21. Collini E, Ferrante C, Bozio R (2005) J Phys Chem B 109:2–5
22. Lepkowicz RS, Cirloganu CM, Fu J, Przhonska OV, Hagan DJ, Van Stryland EW, Bondar MV, Slominsky YL, Kachkovski AD (2005) J Opt Soc Am B 22:2664–2685
23. Zheng S, Leclercq A, Fu J, Beverina L, Padilha LA, Zojer E, Schmidt K, Barlow S, Luo J, Jiang S-H, Jen AK-Y, Yi Y, Shuai Z, Van Stryland EW, Hagan DJ, Brédas J-L, Marder SR (2007) Chem Mater 19:432–442
24. Signorini R, Ferrante C, Pedron D, Zerbetto M, Cecchetto E, Slaviero M, Fortunati I, Collini E, Bozio R, Abbotto A, Beverina L, Pagani GA (2008) J Phys Chem A 112:4224–4234

25. De Boni L, Gaffo L, Misoguti L, Mendonça CR (2006) Chem Phys Lett 419:417–420
26. De Boni L, Correa DS, Pavinatto FJ, dos Santos DS Jr, Mendonça CR (2007) J Chem Phys 126:165102-1–165102-4
27. De Boni L, Piovesan E, Gaffo L, Mendonça CR (2008) J Phys Chem A 112:6803–6807
28. Balu M, Hales J, Hagan DJ, Van Stryland EW (2004) Opt Express 12:3820–3826
29. Balu M, Hales J, Hagan DJ, Van Stryland EW (2005) Opt Express 13:3594–3599
30. Oliveira SL, Corrêa DS, De Boni L, Misoguti L, Zilio SC, Mendonça CR (2006) Appl Phys Lett 88:021911-1–021911-3
31. Corrêa DS, De Boni L, Gonçalves VC, Balogh DT, Mendonça CR (2007) Polymer 48:5303–5307
32. Samoc M, Dalton GT, Gladysz JA, Zheng Q, Velkov Y, Ågren H, Norman P, Humphrey MG (2008) Inorg Chem 47:9946–9957
33. Powell CE, Morrall JP, Ward SA, Cifuentes MP, Notaras EGA, Samoc M, Humphrey MG (2004) J Am Chem Soc 126:12234–12235
34. Samoc M, Morrall JP, Dalton GT, Cifuentes MP, Humphrey MG (2007) Angew Chem Int Ed 46:731–733
35. Coe BJ (1999) Chem Eur J 5:2464–2471
36. Cifuentes MP, Powell CE, Humphrey MG, Heath GA, Samoc M, Luther-Davies B (2001) J Phys Chem A 105:9625–9627
37. Powell CE, Cifuentes MP, Morrall JPL, Stranger R, Humphrey MG, Samoc M, Luther-Davies B, Heath GA (2003) J Am Chem Soc 125:602–610
38. Powell CE, Humphrey MG, Cifuentes MP, Morrall JP, Samoc M, Luther-Davies B (2003) J Phys Chem A 107:11264–11266
39. Cifuentes MP, Powell CE, Morrall JP, McDonagh AM, Lucas NT, Humphrey MG, Samoc M, Houbrechts S, Asselberghs I, Clays K, Persoons A, Isoshima I (2006) J Am Chem Soc 128:10819–10832
40. Samoc M, Gauthier N, Cifuentes MP, Paul F, Lapinte C, Humphrey MG (2006) Angew Chem Int Ed 45:7376–7379
41. Hurst SK, Cifuentes MP, Morrall JPL, Lucas NT, Whittall IR, Humphrey MG, Asselberghs I, Persoons A, Samoc M, Luther-Davies B, Willis AC (2001) Organometallics 20:4664–4675
42. Dalton GT, Cifuentes MP, Petrie S, Stranger R, Humphrey MG, Samoc M (2007) J Am Chem Soc 129:11882–11883
43. Klapshina LG, Grigoryev IS, Lopatina TI, Semenov VV, Domrachev GA, Douglas WE, Bushuk BA, Bushuk SB, Lukianov AYu, Afanas'ev AV, Benfield RE, Korytin AI (2006) New J Chem 30:615–628
44. Luc J, Niziol J, Niechowski M, Sahraoui B, Fillaut J-L, Krupka O (2008) Mol Cryst Liq Cryst 485:990–1001

Luminescent Platinum Compounds: From Molecules to OLEDs

Lisa Murphy and J. A. Gareth Williams

Abstract Around 30 years ago, much of the research into platinum coordination chemistry was being driven either by research into one-dimensional, electrically conducting molecular materials exploiting the stacking interactions of planar complexes, or by the unprecedented success of *cis*-Pt(NH$_3$)$_2$Cl$_2$ (cisplatin) as an anticancer agent. At that time, a number of simple platinum(II) compounds were known to be photoluminescent at low temperature or in the solid state, but almost none in fluid solution at room temperature. Since that time, several families of complexes have been discovered that are brightly luminescent, and a number of investigations have shed light on the factors that govern the luminescence efficiencies of Pt(II) complexes. Over the past decade, such studies have been spurred on by the potential application of triplet-emitting metal complexes as phosphors in organic light-emitting devices (OLEDs), where their ability to trap otherwise wasted triplet states can lead to large gains in efficiency. In this contribution, we take a chemist's perspective of the field, overviewing in the first instance the factors that need to be taken into account in the rational design of highly luminescent platinum(II) complexes, and the background to their use in OLEDs. We then consider in more detail the properties of some individual classes, highlighting work from the past 3 years, and including selected examples of their utility in OLEDs and other applications.

Keywords Electroluminescence, Luminescence, OLEDs, Photochemistry, Platinum

Contents

1 Fluorescence and Phosphorescence: Singlets and Triplets 76
2 Electroluminescence and OLEDs 77

L. Murphy and J.A. Gareth Williams (✉)
Department of Chemistry, University of Durham, Durham, DH1 3LE, UK
e-mail: j.a.g.williams@durham.ac.uk

3 Designing Highly Luminescent Platinum Complexes 79
   3.1 Tuning Excited States to Optimise Luminescence Efficiencies 81
4 PtL₂X₂ Complexes (L = Neutral 2e⁻ Donor Ligand, X = Anionic Ligand) 82
   4.1 X = –C≡N 82
   4.2 X = Organometallic Carbon 85
   4.3 L₂X₂ = Salen and Derivatives 87
5 Pt(N^N⁻)-Based Complexes: 'Pseudo-Cyclometallates' 89
6 Pt(N^C)-Based Complexes (N^C = Bidentate Cyclometallating Ligand) 93
   6.1 The 2-Arylpyridines and Analogues: Versatile Cyclometallating Ligands
        for Colour Tuning of Platinum(II) Complexes 93
   6.2 Excimers and Introduction to WOLEDs 96
   6.3 Multifunctional Complexes 97
7 Complexes with Terdentate Ligands 99
   7.1 Platinum(II) Complexes with N^N^C-Binding Ligands 100
   7.2 Platinum(II) Complexes with N^N^O-Binding Ligands 104
   7.3 Platinum(II) Complexes of N^C^N-Binding Ligands 104
References 109

1 Fluorescence and Phosphorescence: Singlets and Triplets

It was Alexandre-Edmond Becquerel, in his 1867 treatise *La Lumière, ses causes et ses effets*, who first put forward a systematic distinction between fluorescence and phosphorescence. Becquerel designed a phosphoroscope that allowed precise time intervals to elapse between the exposure of a material to light and the observation of the light emitted. He defined fluorescence as emission of light that is immediately extinguished upon removal of the light source, whilst phosphorescence persists for some time after exposure.

The modern and more rigorous distinction is based in quantum theory. Fluorescence is the emission of light that occurs as the result of a spin-allowed electronic transition ($\Delta S = 0$), for example, from the first excited singlet state of an aromatic molecule to the ground state, $S_1 \rightarrow S_0$ [1]. It is normally characterised by a high radiative rate constant k_r^S of the order of 10^8–10^9 s^{-1}, leading to short emission lifetimes in the nanosecond range. Phosphorescence arises from spin-forbidden transitions ($\Delta S \neq 0$), for example, from the first excited triplet state to the ground state, $T_1 \rightarrow S_0$. In this case, radiative rate constants k_r^T are frequently small, around 1 s^{-1} for molecules such as anthracene, leading to natural lifetimes of the order of 1 s [2]. In fact, the phosphorescent process is so slow that faster processes of non-radiative decay of the triplet state normally predominate for such molecules in solution at room temperature, so that phosphorescence is not observed. Only by cooling the sample down to low temperatures and rigidifying it to inhibit such processes is the phosphorescent light observable. The key radiative and non-radiative processes involved are typically represented by a Jabłonski diagram; an example for an organic compound is given in Fig. 1.

In the case of photoluminescence, where the molecule is excited by light, the proportion of triplet and singlet states formed depends upon the relative magnitudes of

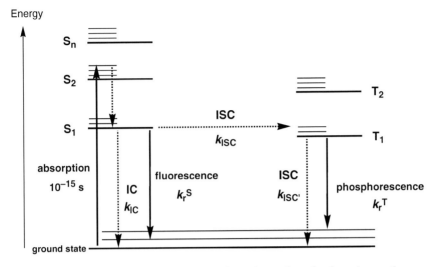

Fig. 1 A simplified Jabłoński diagram for a conjugated organic molecule such as anthracene, illustrating the rate constants for key processes as defined in the text. 'IC' represents internal conversion (an isoenergetic process) followed by vibrational relaxation; similarly 'ISC' represents intersystem crossing followed by vibrational relaxation

k_r^S and k_{ISC}, where k_{ISC} is the rate constant for intersystem crossing from S_1 to T_1. For simple conjugated molecules like naphthalene, anthracene and pyrene, these two parameters have similar values to one another, leading to fluorescence quantum yields of the order of 0.5. In contrast, in cases such as benzophenone, where the difference in energy between S_1 and T_1 is very small, k_{ISC} may be very high such that no significant fluorescence is observed, and the efficiency of triplet formation approaches unity.

Irrespective of the triplet yield, once formed, the radiative rate constant of the triplet state is normally low. If a heavy atom – such as a third row transition metal ion – is brought close to the π-system of the molecule, in such a way that there is significant mixing of the orbitals, then k_r^T can be greatly increased, sometimes by up to 10^6 or more, due to the large spin–orbit coupling associated with the heavy atom which scales with Z^4. Radiative emission can now compete with deactivating processes allowing phosphorescence to be observed at room temperature.

2 Electroluminescence and OLEDs

Organic light-emitting devices (OLEDs) are based on electroluminescence. The energy required to raise the emitting molecule to the excited state is supplied electrically, rather than by the absorption of higher-energy light as it is in photoluminescence. Briefly, a thin layer of the emissive material is sandwiched between two electrodes [3]. Upon application of the electric field, the material is reduced

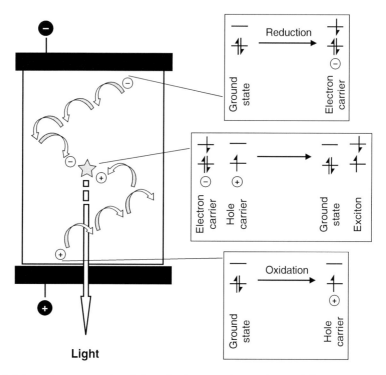

Fig. 2 Schematic representation of electroluminescence in organic materials

at the cathode and oxidised at the anode to give electron and hole carriers, respectively, which migrate under the applied field towards the other electrode. When two such charge carriers of opposite sign meet, they combine to produce one molecule in the ground state and one in the excited state (Fig. 2). In contrast to photoluminescence, both singlet and triplet excited states (also known as excitons) are now formed *directly*, and their proportion is governed primarily by the statistics of charge recombination, leading to a theoretical 1:3 (S:T) ratio [4]. In a purely organic light-emitting device, emission from the triplet states is forbidden, so the efficiency is capped at 25%. This represents a huge wastage of energy, which simply goes into heating up the device.

As noted in Sect. 1, the introduction of a heavy metal ion onto a conjugated organic molecule can greatly accelerate the S ← T process through the influence of spin–orbit coupling. Thus, by incorporating such phosphorescent materials into the emissive layer of the device, emission from the triplet excitons can be promoted, and quantum efficiencies of 100% potentially achieved.

The first metal-based phosphorescent dopant to be investigated as such a 'triplet-harvesting' dopant in an OLED was platinum(II) octaethylporphyrin, **1** [5]. Upon doping this compound into the widely-used EL emissive material Alq$_3$ {tris(8-quinolinato)aluminium, **2**}, 90% of the energy is transferred from the host to the porphyrin, and a device of internal electroluminescence efficiency of 23% (external

EL = 4%) obtained [6]. Given that the triplets are now clearly emitting too, why isn't the device efficiency higher? Part of the reason is the long phosphorescence lifetime of **1** (~80 μs in solution), which results in severe triplet-triplet annihilation, particularly at high current (Eq. 1); in other words, the triplet excited state is so long-lived that it survives long enough to encounter other triplet excitons, and half the energy is lost non-radiatively. Clearly, there is a need for new complexes that emit efficiently but with shorter radiative lifetimes. Ways of achieving this are addressed in the following section.

$$4(^3M^* + {}^3M^*) \rightarrow {}^1M^* + 3 \, {}^3M^* + 4M. \tag{1}$$

1 **2**

3 Designing Highly Luminescent Platinum Complexes

In common with fluorescent molecules, the efficiency of luminescence of metal complexes from triplet excited states, η_r^T, will be favoured by high radiative rate constants k_r^T and low rate constants for non-radiative decay pathways $\sum k_{nr}^T$:

$$\eta_r^T = k_r^T / \left\{ k_r^T + \sum k_{nr}^T \right\}. \tag{2}$$

In photoluminescence, the observed triplet luminescence quantum yield, Φ_{lum}^T, also depends on the efficiency of formation of the triplet state upon absorption of light, in turn determined by the magnitude of the rate constant of intersystem crossing, k_{ISC} (Eq. 3), where k_r^S and $\sum k_{nr}^S$ are the radiative and non-radiative decay rate constants of the singlet state:

$$\Phi_{lum}^T = \eta_r^T \cdot k_{ISC} / \left\{ k_r^S + \sum k_{nr}^S \right\}. \tag{3}$$

The high spin–orbit coupling constant of the platinum nucleus ($\chi = 4{,}481$ cm^{-1}) should facilitate both the S $\rightarrow$ T intersystem crossing process and the T $\rightarrow$ S radiative decay. However, the extent to which it does so in a complex will depend upon the contribution of metal atomic orbitals to the excited state. In many simple Pt(II) complexes with relatively small ligands, the metal's involvement is such that triplet state formation is very fast, of the order of 10^{12} s^{-1} [7]. Since this greatly exceeds typical singlet radiative rate constants of aromatic ligands, emission is then

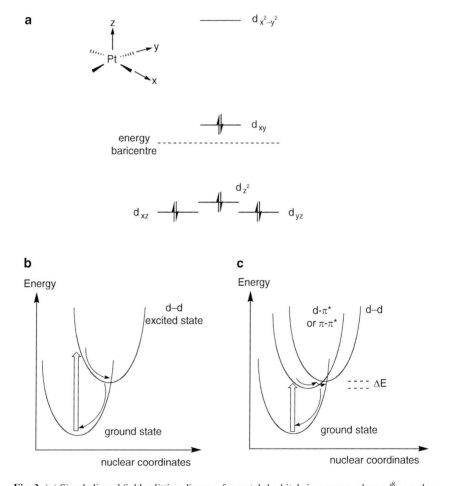

Fig. 3 (a) Simple ligand field splitting diagram for metal d orbitals in a square planar d^8 complex. The z axis is, by convention, perpendicular to the plane of the complex and the M–L bonds lie along the x and y axes. Although the ordering of the lower energy levels depends on the ligand set (e.g. relative importance of σ- and π-effects), the d_{x2-y2} is always the highest. (b) The displaced potential energy surface of the d–d excited state formed by population of the d_{x2-y2} orbital. (c) The d–d state can provide a thermally activated decay pathway for other, lower-lying excited states if ΔE is comparable to kT. *Thick arrows* represent absorption of light; *thin ones* indicate vibrational relaxation and non-radiative decay

observed exclusively from the triplet states. On the other hand, if the excited state is localised on a remote part of a more extended ligand, well removed from the metal centre, or if the pertinent orbitals are orthogonal and hence poorly mixed, ligand-based singlet fluorescence may be observed [8, 9].

Similarly, the radiative rate constant k_r^T is likely to be larger for excited states in which the metal plays a significant role; e.g. MLCT states, or LC states that are significantly perturbed by the metal [10].

Equally important is the need to minimise the rate of non-radiative decay, Σk_{nr}. In the case of square planar platinum(II) complexes, this is often determined primarily by the energy of metal-centred (d–d) states, a point that can be readily understood in terms of the ligand-field splitting diagram of square planar d^8 complexes (Fig. 3a) [11]. The unoccupied d_{x2-y2} orbital is strongly antibonding. Population of this orbital will therefore be accompanied by elongation of Pt–L bonds and severe distortion, which promotes non-radiative decay to the ground state due to the isoenergetic crossing point lying at accessible energies (Fig. 3b). Of course, in many platinum(II) complexes, LC and MLCT states may lie lower in energy than these d–d states, but the latter will still exert a deleterious influence if they are thermally accessible, i.e. if ΔE is comparable to kT (Fig. 3c). Let us take an example. In the simple bipyridyl complex Pt(bpy)Cl$_2$ **3**, the d–d state is the lowest-lying excited state, and the complex does not emit at room temperature [12]. In the analogous complex of the ester-substituted bipyridine, Pt{3,3'-(MeCO$_2$)$_2$-bpy}Cl$_2$ **4**, the lowest-lying excited state is now the MLCT state, but the d–d state lies only a little higher in energy, offering the former a non-radiative decay pathway and also rendering this complex non-emissive at room temperature [13].

3.1 Tuning Excited States to Optimise Luminescence Efficiencies

The above discussion leads naturally to the conclusion that the most emissive platinum(II) complexes will be those in which (1) the lowest-energy excited state is *not* a metal-centred d–d state but rather an LC or CT state, and (2) where the energy gap between the lowest excited state and the d–d state is sufficiently large as to prevent thermally activated decay via the latter.

For example, platinum(II) porphyrins are perhaps the oldest known class of luminescent Pt(II) complexes. They emit deep into the red region (e.g. λ_{max} = 641 nm, Φ_{lum} = 0.6 for **1** [14]). Clearly the emissive, porphyrin-based π–π^* state is sufficiently low in energy that activation to the d–d state is ruled out at room temperature. However, relying on the low-energy of the emissive state to disfavour non-radiative decay is clearly not an option if higher-energy blue- or green-emitting complexes are sought! In such cases, it is necessary to work on ensuring that the d–d states in the complex are kept at high energies [11]. In practice, this means that strong-field ligands need to be present in the coordination sphere of the metal ion.

For example, in contrast to Pt(bpy)Cl$_2$ mentioned above, complexes of the form Pt(bpy)(–C≡C–Ar)$_2$ are normally quite strongly emissive in solution at room temperature [15]. The difference can be attributed – at least in part – to the stronger ligand-field associated with the acetylide compared to the chloride co-ligands. Similarly, cyclometallated ligands, particularly those offering the metal ion a synergistic combination of a π-accepting heterocycle (e.g. pyridine) with the strongly σ-donating cyclometallated carbon, lead to a strong ligand field. This accounts, for example, for [Pt(N^C-ppy)(N^N-bpy)]$^+$ being luminescent at room temperature, in contrast to [Pt(N^N-bpy)$_2$]$^{2+}$, which is not (ppyH = 2-phenylpyridine) [16]. A more comprehensive discussion of these principles is provided in a recent contribution to the sister volume of this series [11].

In this chapter, we shall overview some of the main classes of platinum(II) complexes that are luminescent under ambient conditions, highlighting those which have been shown to be potentially useful OLED dopants [17]. The emphasis is on examples from the past 3 years, although naturally not to the exclusion of important and informative earlier original examples.

4 PtL$_2$X$_2$ Complexes (L = Neutral 2e$^-$ Donor Ligand, X = Anionic Ligand)

Many of the complexes which fall into this class are Pt(N^N)-based compounds, where N^N is a bidentate diimine ligand such as bipyridine or phenanthroline and derivatives thereof. As noted in Sect. 3, those where X = Cl are rarely significantly emissive, because the *d-d* state is either the lowest-energy excited state or is thermally accessible from a lower-lying CT or LC state [13].

4.1 X = –C≡N

The substitution of the halides by strong-field cyanides (X = –C≡N) lowers the energy of the highest-occupied metal-centred orbitals and raises the vacant d_{x2-y2} orbital, such that Pt(N^N)(CN)$_2$ complexes normally emit from π–π* states. Many of these complexes also form emissive excimers [18–20]. For example, Pt(dpphen)(CN)$_2$ **5** displays green emission centred at 520 nm in dilute CH$_2$Cl$_2$ (10^{-5} M), but increasing the concentration leads to the appearance of a red emission band, centred at 615 nm, which grows in at the expense of the green band with an identical excitation spectrum [19]. The red emission is attributed to an excimer analogous to well-known examples amongst planar organic compounds such as pyrene, and formed from the bimolecular interaction of an excited monomer with a ground-state monomer [21].

5   **6**

Very recently, Kato and colleagues prepared Pt(dC$_n$bpy)(CN)$_2$ **6** complexes bearing long alkyl chains, where dC$_n$bpy represents bipyridine substituted at the 4 and 4′ positions by –C$_9$H$_{19}$ ($n = 9$) or –C$_{11}$H$_{23}$ ($n = 11$) [22]. These compounds could be prepared by treatment of the respective ligands with platinum dicyanide in a mixture of DMF and aqueous ammonia. The long chains render the complexes soluble in a wider range of solvents than the parent systems, and to higher concentrations, which has allowed the solvent dependence of excimer formation to be investigated. In polar solvents such as methanol and acetonitrile, exclusively monomer emission is observed over the concentration range 10^{-5} to 10^{-3} M. Over the same range in chloroform, the emission switches from monomer to excimer, whilst in toluene solution, the excimer is observed even at 10^{-5} M. Evidently, the reorganisation of the molecules to form the excimer is favoured by solvents of low polarity. The difference is strikingly revealed by the change in the emission spectrum of a 5×10^{-5} M solution of the $n = 9$ complex in different ratios of CHCl$_3$/toluene (Fig. 4).

In an unusual recent development, Kunkely and Vogler have investigated the excited state properties of Pt(COT)(CN)$_2$ **7** (COT = cyclo-octatetraene) [23]. The COT ligand can be thought of as an analogue of a diimine, in that it has relatively low-energy π* orbitals. Indeed, the complex in the solid state displays a long-wavelength absorption maximum centred at 430 nm and intense orange emission at 572 nm, attributed to the singlet and triplet MLCT {Pt^{II} → π*(C$_8$H$_8$)} states respectively (Fig. 5). Unfortunately, in solution, the complex is of limited stability and degrades with release of COT.

In addition to excimer formation, square planar platinum(II) complexes are frequently able to participate in ground-state intermolecular interactions through their axial positions. In some cases, this leads to oligomerisation through overlap of the d_{z^2} orbitals that are orthogonal to the plane of the complex [24, 25]. The combination of isonitriles and cyanides as ligands for Pt(II) has been extensively studied by Mann and co-workers in their work on vapochromic sensors [26, 27]. The intermolecular interactions in double salts of the type [Pt(CNR)$_4$][Pt(CN)$_4$] and the related neutral complexes [Pt(CNR)$_2$(CN)$_2$]$_2$ are influenced by the presence of volatile organic solvents, leading to characteristic changes in colour [28].

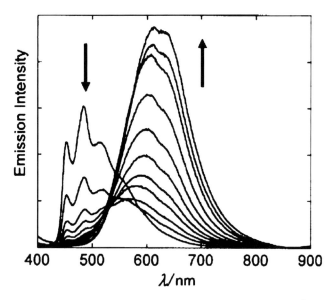

Fig. 4 Change in the emission spectrum of [Pt(dC$_9$bpy)(CN)$_2$] (**6**, $n = 9$) (5×10^{-5} M; $\lambda_{ex} = 320$ nm) in different ratios of CHCl$_3$/toluene, from 100% CHCl$_3$ to 100% toluene in 10% steps [22]

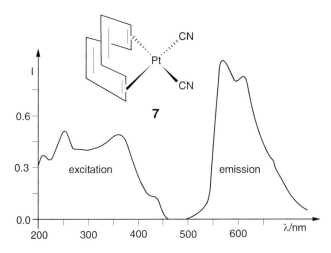

Fig. 5 Excitation ($\lambda_{em} = 572$ nm) and emission ($\lambda_{ex} = 400$ nm) spectra of solid Pt(COT)(CN)$_2$ **7** at room temperature. Reprinted from [23] with permission from Elsevier

Recently, Wang and Che have described luminescent micro- and nano-wires that form by self-assembled aggregation of [Pt(CN^{t}Bu)$_2$(CN)$_2$] directed by Pt⋯Pt interactions [29]. This complex itself is non-emissive in acetonitrile solution at room temperature, but displays bright green luminescence from the rod-like crystalline form. The corresponding X-ray structure reveals infinite linear stacks

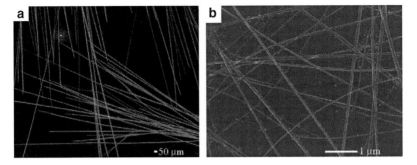

Fig. 6 (a) Emission microscopy image of luminescent [Pt(CN^{t}Bu)$_2$(CN)$_2$] wires formed on a silicon substrate by slowly evaporating a 10^{-4} M solution in MeCN. (b) Scanning electron micrograph of thinner wires of the same compound formed by injecting 50 mL of a 9.6 × 10^{-3} M solution in MeCN into 10 mL of water, followed by transfer onto a silicon substrate. Note the different magnification in the two images. Reprinted from [29] with permission from Wiley

comprised of pairs of complexes, with a torsion angle of 125.20° between them about the stack axis, and close Pt⋯Pt distances of 3.354(1) Å. Evaporation of an acetonitrile solution in air on a silicon substrate led to luminescent wires of diameter approximately 25 μm and lengths as great as 1,000 μm or more, which could be readily visualised by emission microscopy (Fig. 6a). Wires of smaller diameter could be prepared by prior dilution of the acetonitrile solution with water; an example visualised by scanning electron microscopy is shown in Fig. 6b. The luminescence emitted by the wires (λ_{max} = 562 nm at 298 K) resolves out into two components at 77 K (λ_{max} = 545 and 597 nm). The former is attributed to the dimeric Pt$_2$ ^{3}[5dσ* → 6pσ] excited state and the latter to trimeric and tetrameric species, an assignment supported by recent density functional theory (DFT) calculations [30].

4.2 X = Organometallic Carbon

Organometallic derivatives of the form Pt(N^N)(–Ar)$_2$, incorporating Pt–aryl bonds, have also been studied. They are accessible by reaction of the diimine ligands (e.g. bpy or phen) with Pt(COD)(aryl)$_2$ precursors, in turn formed from Pt(COD)Cl$_2$ and the corresponding aryl-lithium reagent [31]. The bis-mesityl complex **8** emits at 660 nm in toluene solution at room temperature from a ^{3}MLCT state [32]. Clearly, the strongly σ-donating aryl groups serve to lower the energy of the excited state. The mesityl complexes are unusual amongst square planar platinum(II) compounds in that they can be reversibly oxidised to remarkably persistent d^7 Pt(III) species, an effect attributed to the effective protection of the axial positions by the two mesityl groups [31].

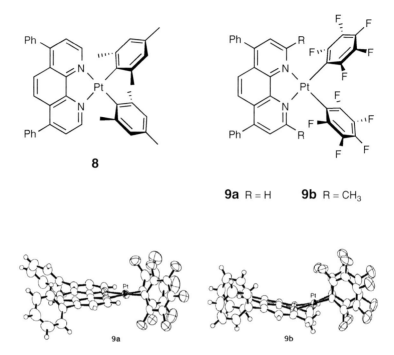

Fig. 7 Molecular structures of **9a** and **9b** in the solid state from X-ray diffraction analysis [33]

More recently, related compounds containing perfluorinated aryl rings have been prepared, e.g. **9** [33]. Like the mesityl systems, the aryl rings are twisted by about 70° with respect to the PtN$_2$C$_2$ square plane in the solid state structures (Fig. 7). The emission of Pt(dpphen)(Ar^F)$_2$ **9a** (λ_{max} = 515 nm in CH$_2$Cl$_2$) is substantially blue-shifted compared to the mesityl compound **8**, which can be readily rationalised in terms of the electron-withdrawing influence of the fluorine substituents in lowering the energy of the metal-centred HOMO. No emission could be detected from the analogous complex incorporating methyl substituents at the 2 and 9 positions of dpphen, **9b**. The X-ray structure of this complex reveals a bent structure (Fig. 7) and slightly longer Pt–N bonds owing to steric hindrance of the methyl groups, both of which features will lead to a weaker ligand field at the metal and to a less rigid structure, promoting non-radiative decay. The fluorine substituents aid sublimation, and a multi-layer OLED incorporating 6% by mass of **9a** within a CBP host emissive layer has been prepared by thermal deposition; it displays an external quantum efficiency of 2.1% at 100 cd m^{-2}.

Fekl and co-workers have recently reported the first examples of diimine-coordinated platinum complexes with 9,10-dihydroplatinaanthracene co-ligands [34]. These compounds are metallacycles: the Pt ion replaces one carbon atom within the carbocyclic system (anthracene). Reaction of the new organometallic oligomeric platinum precursor {[H$_2$C–(C$_6$H$_4$)$_2$]Pt(SEt$_2$)}$_n$ (n = 2,3) with diimines

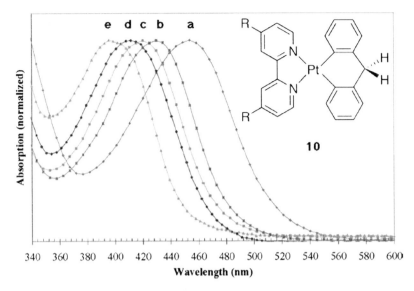

Fig. 8 Normalised charge-transfer-to-diimine absorption bands for metallacycles **10a–e** in MeCN at room temperature. Identity of R: a = Cl, b = H, c = t-Bu, d = MeO, e = NMe$_2$. Reprinted from [34] with permission from ACS

such as bipyridines, phenanthrolines and biquinoline provides easy access to a whole range of new complexes incorporating the metallacycle. The lowest-energy absorption bands of the bipyridyl series **10a–e** increase in energy as the diimine becomes more electron-rich, in line with charge-transfer-to-diimine character to the lowest spin-allowed excited state (Fig. 8). This is supported by electrochemistry and DFT calculations, which suggest that the HOMO is based primarily on the platinacycle. Although none of the compounds prepared to date display detectable emission in solution at room temperature, they emit brightly in the solid state, showing a trend in emission energies that is largely consistent with that in absorption.

4.3 L_2X_2 = Salen and Derivatives

Oxygen ligands on their own are generally not suitable for forming stable complexes with the soft Pt(II) ion, but in combination with other ligands such as diimines, complexes with high thermodynamic stability can be obtained. Shagisultanova reported that the well-known tetradentate ligand, N,N'-bis(salicylidene)-1,2-ethylenediamine or salen, forms a highly luminescent complex with Pt(II), **11** [35]. The emissive excited state is probably one of d(Pt)/π(O) → π*(Schiff base) character. Che and co-workers investigated this complex and its tetramethylethylenediamine analogue as sublimable, green-emitting OLED dopants [36]. At the optimised dopant concentration of 4 wt%, a maximum external quantum efficiency of 11% was achieved at 8.5 mA cm^{-2}. The same team has also explored complexes of related ligands based on dpphen or tbbpy appended with ortho-phenoxy groups, which prove to be superior in terms of solution quantum yields; e.g. $\Phi = 0.6$ for **12** in DCM compared to 0.19 for **11** in MeCN [37].

11 R = –H
13 R = –OCH3

12

Wong and co-workers have recently described the result of intriguing intermolecular packing effects in a dimethoxy-substituted salen complex **13** [38, 39]. For example, different solvates have been crystallised, **13**.H$_2$O and **13**.DMF, which are orange and red respectively. The very different packing arrangements in the two crystals are shown in Fig. 9a, b. Meanwhile, the self-assembly of **13** with K$_2$Pt(CN)$_4$ or K$_2$Pd(CN)$_4$ leads to compounds [K$_2$(**13**)$_2$M(CN)$_4$] (M = Pd or Pt, isostructural; Fig. 9c). K–O coordination in the crystals of these compounds brings the **13** units

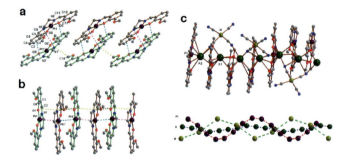

Fig. 9 Packing of the molecules in the crystals of (**a**) **13**.H$_2$O and (**b**) **13**.DMF, from X-ray diffraction analysis. Hydrogen atoms are omitted for clarity. (**c**) The packing and one-dimensional helical chains observed in the structures of [K$_2$(**13**)$_2$ M(CN)$_4$] (M = Pd or Pt). Reprinted from [38] with permission from Elsevier

closer together than in the two solvates, probably accounting for the greater red component in the solid-state emission spectrum of the former compounds [38].

5 Pt(N^N⁻)-Based Complexes: 'Pseudo-Cyclometallates'

Whereas pyridines and quinolines necessarily bind to metal ions as neutral two-electron 'L' donors, heterocycles such as pyrazole, indazole, imidazole and triazole can do so either via a neutral nitrogen atom or through deprotonation of a 'pyrrolic' nitrogen, as does pyrrole itself. Bidentate ligands that incorporate each type of ring, for example 3-(2-pyridyl)pyrazole, can thus frequently bind either as N^N or N^N⁻ ligands (Fig. 10). In the latter case, the coordination offered to the metal is essentially analogous to that of a cyclometallating ligand such as 2-phenylpyridine (ppy), and the resulting complexes could be construed as pseudo-cyclometallates, in which the N⁻ atom resembles a metallated carbon. Since such ligands should exert a strong ligand field, a beneficial effect on luminescence efficiencies might be anticipated.

Chi and co-workers have explored the chemistry of platinum(II) with a series of 3-(2-pyridyl)pyrazole, 3-(pyrazyl)pyrazole and 3-(2-pyridyl)triazole ligands, and have noted that their luminescence quantum yields in solution vary widely according to the identity of substituents on the rings [40]. For example, the *tert*-butyl-substituted complexes **14** and **15** are amongst the most intensely luminescent platinum(II) complexes known ($\Phi = 0.82$ and 0.86; $\tau = 2.4$ and 2.1 μs respectively). In contrast, the analogues **16a** and **16b**, incorporating fluorinated substituents, and the triazole systems **17a** and **17b** display only very weak and short-lived emission under the same conditions ($\Phi < 10^{-3}$; $\tau < 10$ ns), possibly due to severe self-quenching. Nevertheless, bright emission is observed from thin solid films of these compounds, probably due to aggregation to give a Pt$_2$ d$\sigma^* \to \pi^*$(N^N) excited state. This assignment is supported by the short Pt⋯Pt contacts (3.442 Å) in the

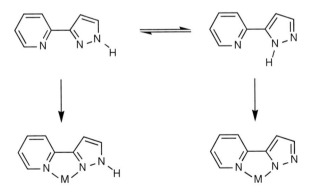

Fig. 10 The tautomeric forms of 3-(2-pyridyl)-pyrazole can allow it to coordinate to a metal M as either a neutral N^N-binding ligand, or as an anionic N^N⁻ ligand, analogous to a cyclometallating N^C ligand

crystal structure of **16b**, compared to a displaced arrangement of the molecules in the structure of **15**, where Pt···Pt distances are too long for significant interaction of d_{z^2} orbitals. Both types of complexes have been successfully incorporated into electroluminescent devices by doping into a CBP host; e.g. an external quantum efficiency of 6% was observed in such a device employing **16a**.

The change from N^N to N^N⁻ and the beneficial effect this can have on luminescence efficiency is dramatically illustrated by the platinum(II) complexes of a pair of isoquinolyl–indazole ligands also investigated by Chi [41]. The complexes in which these ligands are bound as neutral N^N ligands, **18** and **19**, are non-emissive in solution, just like Pt(bpy)Cl$_2$. However, upon reaction with anionic ligands such as picolinate or a pyridylpyrazolate, deprotonation of the indazolic N–H occurs to give N^N⁻-bound complexes, e.g. **20** and **21**. These complexes are luminescent in solution at room temperature, a change that can be attributed to the increase in ligand field strength that accompanies deprotonation. The 1-linked isomers are consistently better emitters than the 3-linked isomers (e.g. $\Phi_{\text{lum}} =$ 0.64 and 0.035, $\tau = 8.2$ and 0.85 µs for **20** and **21** respectively, in CH$_2$Cl$_2$ at room temperature), which seems to be related primarily to faster non-radiative decay in the latter.

Pyrroles can function in a similar way to the deprotonated pyrazoles of the above systems, when incorporated into chelating ligands reminiscent of salen, as demonstrated by Che and co-workers [42]. They prepared the complexes **22–24** which display the emission spectra shown in Fig. 11. Solution quantum yields of **22** and **23** are of the order of 0.10 (in MeCN at 298 K), while the shift to lower energy in the phenylene-based system **24** – associated with the more extended π-conjugated system – is accompanied by reduction in Φ_{lum} to 0.001. OLEDs incorporating **22** and **23** doped into a CBP emissive layer have been fabricated. In the case of **22**, the spectral output is dependent on dopant concentration, λ_{max} increasing from 580 to 620 nm on going from 0.8 to 6.0 wt%. This change is attributed to excimer or oligomer formation at the higher loadings. Notably, the colour coordinates (0.62, 0.38) of the 6% system are close to those for pure red light (0.65, 0.35). No such concentration dependence is observed for **23**, probably because the methyl groups inhibit the close approach of the molecules that is necessary for the oligomer or excimer to form.

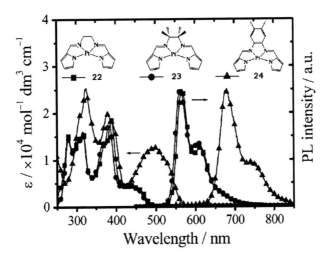

Fig. 11 Structural formulae of **22–24** and their absorption and photoluminescence spectra in MeCN at 298 K. Reprinted from [42] with permission from RSC

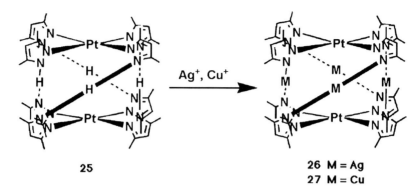

Fig. 12 The synthesis of **26** and **27** by treatment of **25** with Ag$^+$ or Cu$^+$ respectively

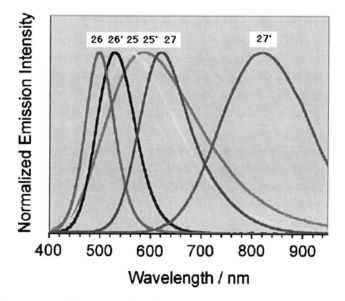

Fig. 13 Normalised emission spectra of **25**, **26** and **27** in the solid state (no prime) and in CH$_2$Cl$_2$ at 295 K (prime), λ_{ex} = 266 nm. Reprinted from [43] with permission from ACS

Platinum-containing heteropolynuclear complexes based on deprotonated pyrazoles have recently been reported by Umakoshi et al. [43]. These intriguing compounds have formulae [Pt$_2$M$_4$(μ-Me$_2$pz)$_8$] (M = Ag or Cu; **26** and **27** respectively), and are prepared from the corresponding protonated systems (M = H, **25**) (Fig. 12) {Me$_2$pz = 3,5-dimethylpyrazolate). The compounds **25**, **26** and **27** display yellow, blue and orange luminescence (Φ_{lum} = 0.06, 0.85 and 0.28) respectively, in the solid state at 295 K. They also emit in solution, although substantially more weakly, probably because of significant structural changes accompanying the formation of the excited state, as suggested by the large red-shifts in solution compared to the solid state (Fig. 13). DFT calculations highlight the importance of metal-metal bonding in the frontier orbitals and indicate that the microsecond emission of

6 Pt(N^C)-Based Complexes (N^C = Bidentate Cyclometallating Ligand)

6.1 The 2-Arylpyridines and Analogues: Versatile Cyclometallating Ligands for Colour Tuning of Platinum(II) Complexes

The bis-cyclometallated complex Pt(N^C-thpy)$_2$ **28** (thpyH = 2-thien-2-ylpyridine), reported by von Zelewsky in the mid-1980s [44], was one of the first clear-cut examples of a simple platinum(II) complex that phosphoresces in fluid solution at room temperature (λ_{max} = 578 nm, τ = 2.2 µs in a PrCN/MeCN mixture) [45]. Barigelletti et al. showed, by means of a study of the temperature dependence of emission, that the influence of cyclometallation is to displace the d–d state to about 3,700 cm^{-1} above the emissive state, ensuring that thermally-activated non-radiative decay is blocked off [46]. Although not thermally stable and hence not suitable for vacuum deposition methods of OLED fabrication, this complex and the trimethylsilyl derivative **29** has been incorporated into devices by spin-casting with TPD as a host material, and an EL efficiency of 11.5% has been achieved with **29** [47, 48].

28 R = H

29 R = SiMe$_3$

The formation of such bis-cyclometallated complexes incorporating two such N^C ligands is not trivial, requiring the use of lithiated ligands, the formation of which is intolerant of many functional groups. In fact, complexes with just one such ligand and a second, coordinating (as opposed to metallated) bidentate ligand are excellent alternatives and are normally more readily prepared. For example, it is often possible to treat the N^C ligand with one equivalent of K$_2$PtCl$_4$ to generate a dichloro bridged dimer, (N^C)Pt(μ-Cl)$_2$Pt(N^C), which can then be cleaved with a variety of monodentate or bidentate ligands [49]. The dimers are usually formed readily in mixed solvent systems such as ethoxyethanol/water at 80°C. Reaction times are usually of the order of 24 h, although Ghedini and co-workers have recently reported that microwave irradiation can greatly speed up the reaction, provided that the temperature does not rise beyond about 65°C, which seems to induce degradation of the platinum salt [50].

Treatment with bidentate diamines, diimines or bis(pyrazolyl)borates gives [Pt(N^C)(N^N)]$^+$ monometallic complexes [16, 51]; bidentate anionic ligands L^X such as β-diketonates give neutral Pt(N^C)(O^O) [52]. Unidentate ligands can also be introduced to give, e.g. [Pt(N^C)Cl$_2$]$^-$, Pt(N^C)(CO)Cl, Pt(N^C)(CO)SR [16, 53, 54]. In fact, the choice of the other ligands is crucial to luminescence efficiency. For example, the ligand field strength in [Pt(N^C)Cl$_2$]$^-$ is inadequate to inhibit non-radiative decay via the d–d state, whereas diimine complexes such as [Pt(N^C)(bpy)]$^+$ are significantly emissive at room temperature. In the latter case, the bpy also influences the nature of the excited state, since the LUMO is largely localised on this ligand, leading to a d(Pt) → π*(N^N) excited state [16].

The β-diketonate class Pt(N^C)(O^O) (e.g. O^O = acac) have proved to be very popular, as the O^O ligand is easily introduced and normally does not influence the excited state energy significantly, so that the properties are essentially those associated with the N^C ligand. Thompson and co-workers carried out a landmark study of such systems with acac or its *tert*-butyl analogue dipivaloylmethane (dpm), in which 23 different N^C-coordinating ligands were investigated [52]. This study revealed that the d(Pt)/π(N^C) → π*(N^C) emission of these systems can be tuned over a wide range by relatively simple structural modification of the N^C ligand, in a way that is quite similar to that achieved over the past decade with N^C-based iridium(III) complexes. Briefly, electron-withdrawing substituents in the aryl ring and electron-donating substituents in the pyridine ring lower the HOMO and raise the LUMO respectively, leading to a shift to the blue. In contrast, extending the conjugation in the heterocycle through the use of quinolines, or using benzothiazole in place of phenyl as the cyclometallating ring, shifts the emission to the red. The extremes are represented by **30** and **31** which, at 77 K, have emission maxima of 440 and 600 nm respectively. Although many of the complexes also emit quite strongly at room temperature, this is not true for all; e.g. for **30**, $\Phi_{lum} < 10^{-3}$, possibly due to a small energy gap between the emissive and higher-lying d-d state.

30

31

A series of related complexes incorporating styryl pendants **32a–d** (Fig. 14) have been investigated by Guerchais and co-workers, which display some rather intriguing properties arising from the possibility of *trans–cis* isomerisation of the

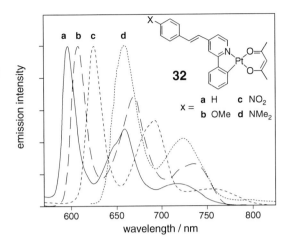

Fig. 14 The structures of the styryl-appended complexes **32a–d** and the long wavelength region of their emission spectra, where the *E* isomers emit, in an EPA glass at 77 K; $\lambda_{ex} = 400$ nm

C=C bond [55]. Emission is weak at room temperature, due to competitive C=C photoisomerisation, as confirmed by ^{1}H NMR spectroscopy. Little effect of the pendant on the emission energy is observed under these conditions, suggesting that the emissive state at room temperature is decoupled from the styryl group, perhaps indicative of a half-twisted –CH=CH– unit.

In contrast, in a rigid frozen glass at 77 K, two sets of bands with similar, well-defined vibrational progressions are observed, one in the region 460–560 nm, and the second set at substantially lower energy in the range 570–800 nm. These are assigned to the *cis* (*Z*) and *trans* (*E*) isomers respectively, evidence for which includes the change in their relative intensities according to the excitation wavelength. The former decrease in relative intensity as the excitation wavelength is increased from 350 to 430 nm, which correlates with the longer-wavelength absorption by the *E* isomer. The energy of the *E* bands is influenced by the substituent, as shown in Fig. 14, which reveals that the emission can be pushed right out to the far red by using an amine pendant [55].

In some complexes, the energy of the L^X ligand in Pt(N^C)(L^X) may be lower than that associated with the Pt(N^C) unit, in which case emission is likely to emanate from an L^X-based excited state. An example is provided by 8-quinolinol (the anion of 8-hydroxyquinoline, HQ) as a ligand, e.g. complex **33**. Here weak emission deep into the red is observed in solution at room temperature, which can

33

be further shifted to the NIR upon introducing substituents into the Q ligand [56, 57]. The emitting state has ILCT character, with a HOMO localised on the phenolate and LUMO on the pyridine ring of the same ligand, just as for the homoleptic complex Pt(quinolinol)$_2$ complexes investigated 30 years ago [58]. Despite the low quantum yield in solution (<0.01) and in powdered form (0.003), complex **33** has been doped into the emissive CBP layer of a multilayer OLED, giving a device displaying saturated red output (CIE coordinates 0.71, 0.28), albeit of efficiency that is too low to be practicable [59].

6.2 Excimers and Introduction to WOLEDs

As noted in Sect. 4.1, sterically unencumbered platinum(II) complexes may sometimes form excimers and/or aggregates, which emit at lower energy than the isolated monomeric molecules. In terms of applications to OLEDs, the combination of monomer and excimer emission from a single metal complex as dopant may offer an intriguing way forward in the development of white light-emitting devices (WOLEDs) [60]. Provisionally these have vast potential as efficient replacements of incandescent light bulbs and fluorescent strip lighting: around 22% of energy consumption in the U.S. is estimated to go on lighting [61]. White light is normally generated by using three separate emitters – red, green and blue [62]. If all three are combined within a single emitting layer, it can be difficult to control energy transfer from the higher energy blue to the green to red dye: energy tends to be 'short-circuited' to the lowest energy emitter. Segregation of the emitters into different layers can circumvent this problem, but this requires a more elaborate device architecture [63]. Thompson and Forrest put forward the novel approach of making use of a single phosphorescent dopant emitting simultaneously from monomer (blue–green) and excimer (green–red) states [60]. Clearly, Pt(II) complexes are potentially excellent candidates for this.

The trick is to achieve the right balance of excimer and monomer at the right sort of doping level, and this can be controlled to some extent through sterics. For example, photoluminescence spectra of a CBP host doped with increasing concentrations of Pt(N^C-F$_2$ppy)(acac) demonstrate that, at a low doping level of 1.5%, undesired residual emission from the host is observed in addition to that of the monomeric Pt complexes [64]. As the doping level is increased, the excimer emission rapidly appears and dominates at levels >8%, which is too low to achieve colour-balanced emission in a practicable device. If the complex is made more bulky to inhibit partially excimer formation, the onset of excimer emission can be limited to higher concentrations. The dpm (dpmH = dipivaloylmethane) derivative is *too* bulky: almost exclusively monomeric emission is observed even at 20% doping level. The mhpt complex, on the other hand, offers just the right level of steric bulk for balanced emission at 10% doping, and this behaviour has been successfully extrapolated to an electroluminescent device (Fig. 15) (mhptH = 6-methyl-2,4-heptanedione). The CIE coordinates of this device (0.36, 0.44) are

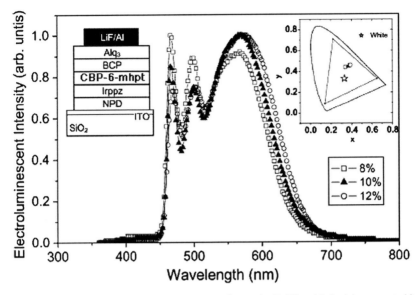

Fig. 15 Electroluminescence spectra and CIE coordinates for Pt(F2ppy)(mhpt) incorporated into the emissive layer of the multilayer device represented schematically in the inset, at the % dopings by mass indicated. Reproduced from [64] with permission from RSC

close to those for white light (0.33, 0.33), with a maximal efficiency of 3.3 ± 3 photons/electron at 0.5 cd m^{-2} [64].

6.3 Multifunctional Complexes

In most devices, the emissive layer is sandwiched between hole- and electron-transporting layers, in order to achieve balanced injection and transport of holes and electrons to optimise charge recombination within the EL layer. Wong et al. have been investigating multifunctional Pt emitters in which oxadiazole and triarylamine units are covalently linked to a Pt(ppy)(acac) emissive centre, to act as electron- and hole-transporting units respectively, e.g. **34** [65]. These sublimable compounds have been used to prepare neat-emissive-layer devices by vacuum evaporation between an ITO/CuPC anode and Ca/Al cathode. Molecules such as **34** display fluorescence from the arylamine as well as Pt-based phosphorescence, both in solution and when doped in the CBP emissive host. The ratio of fluorescence to phosphorescence can be controlled according to either the applied voltage (the blue colour intensity increases relative to the orange as the voltage increases), or the concentration of the dopant (intermolecular quenching of triplet excitons at high loadings leads to an increase in the proportion of blue component). Under appropriate conditions, the combination of the two contributions can be tuned to give WOLEDs, offering an alternative to the monomer-excimer approach [66].

34

35
a R = Ph₂N–
b R = (p-Me-C₆H₄)₂N–   c R = H–

Structurally analogous molecules incorporating fluorene instead of the oxadiazole have been prepared by the same group **35a–c** [67]. Again, the introduction of diarylamino substituents onto the fluorene leads to ILCT singlet fluorescence in addition to the Pt-based phosphorescence (Fig. 16). Notably, the 77 K phosphorescence lifetimes of the substituted systems are significantly longer than the unsubstituted analogue, in line with a smaller contribution of metal character in the excited state of these more extended systems, even though the triplet excited state energies are essentially unaffected.

Finally, we note that Marder and co-workers have recently reported on the covalent incorporation of Pt(N^C)(O^O) complexes into a polymer by ring opening metathesis copolymerisation of norbornene-appended complexes, e.g. **36**, with a norbornene-appended carbazole **37** [68]. Such an approach could provide

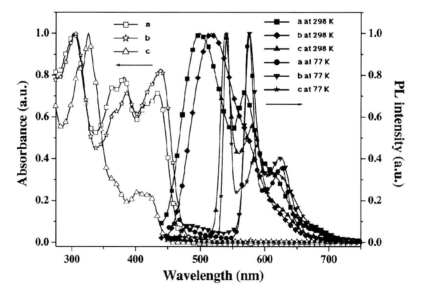

Fig. 16 Absorption and photoluminescence spectra of **35a–c** in CH₂Cl₂. Reprinted from [67] with permission from ACS

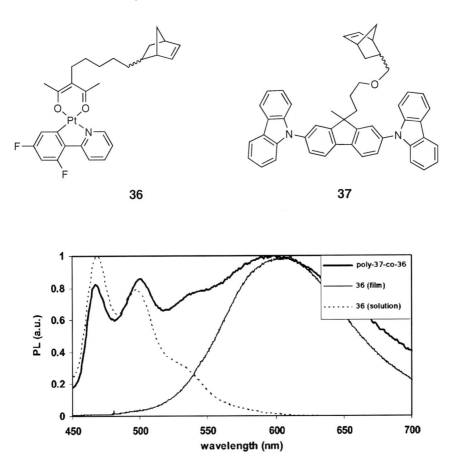

Fig. 17 Photoluminescence spectrum of a film of copolymer **37-co-36**, and corresponding spectra of monomer **36** as a thin film and in dilute solution in THF. Reprinted from [68] with permission from ACS

a solution to some of the problems associated with phase separation in blends of small-molecule phosphors and polymers. The photoluminescence spectra of the films formed in this way display both monomeric and excimer/aggregate emission (Fig. 17), as also proves to be the case in the electroluminescent devices.

7 Complexes with Terdentate Ligands

The d–d excited states of d^8 complexes comprising bidentate ligands are unstable with respect to distortion away from the D_{4h} symmetry of the ground state towards a D_{2d} symmetry, in which the plane of the bidentate ligand is twisted relative to that containing the other two ligands [69]. As highlighted in Fig. 3, excited state distortion is associated with increased non-radiative decay, with an adverse effect on luminescence quantum yields. Terdentate ligands analogous to bpy and ppy, able to bind via three rings instead of two in a planar conformation, should in principle inhibit such distortion owing to a more rigid binding [70]. However, the bite angle associated with many such ligands, e.g. terpyridine (tpy), is not ideal for larger metal ions, leading to a weaker ligand field compared to that which might be anticipated. Thus, [Pt(tpy)Cl]$^+$, for example, is not emissive at room temperature, because the *d–d* state is low in energy [71]. Again, however, the introduction of stronger-field co-ligands in place of Cl can circumvent this problem, and a wide range of emissive complexes based on the [Pt(tpy)(–C≡C–R)]$^+$ structure have been investigated. The reader is directed to recent reviews on such systems by Castellano et al. and by Wong and Yam [15, 72]. In this contribution, our focus is on charge-neutral systems, which are more relevant to the topic of OLEDs. A variety of charge-neutral Pt(II) complexes have been obtained by employing cyclometallating terdentate ligands in combination with an anionic fourth ligand, selected examples of which are discussed in the sections that follow.

7.1 Platinum(II) Complexes with N^N^C-Binding Ligands

Che and co-workers have pioneered much of the work on cyclometallated Pt(II) complexes with the N^N^C-coordinating ligand, 6-phenylbipyridine (phbpyH) and derivatives [73]. Of particular interest is their comprehensive study of around 30 such complexes incorporating σ-alkynyl ligands in the fourth coordination site of the metal ion, [Pt(N^N^C)(–C≡C–R)] [74]. The majority of the complexes are emissive in solution at room temperature, giving featureless spectra ranging from green to red. The emission in most of the systems is attributed to a ^{3}MLCT (d$_{Pt}$ → π*$_{NNC}$) state, an assignment that is supported by the influence of structural modification of both the N^N^C ligand and the acetylide on the emission properties. For example, for a series of *p*-substituted aryl acetylides **38**, the emission maximum shifts to higher energy as the substituent X becomes more electron-withdrawing, and the metal-based HOMO is lowered in energy (Fig. 18). A good negative linear correlation of log Σk_{nr} with the emission energy illustrates that the energy gap law is obeyed within this series (Fig. 18, inset). Meanwhile, introduction of electron-donating *tert*-butyl groups into the 4-positions of the pyridyl rings of the N^N^C ligand shifts the emission energy to the blue (**39**, R = *t*-Bu vs. R = H), whilst an ester at this position (**39**, R = –CO$_2$Et) has the opposite effect, reflecting the contrasting influence of these substituents in raising and lowering the LUMO energy respectively. Substituents in the cyclometallating phenyl ring were found to have rather little influence, but changing from phenyl to thienyl or furanyl rings (**40**, X = S or O), led to a substantial red shift in emission, which can be attributed to the more electron-rich nature of these heterocycles. The complexes are thermally stable and a number of them have been incorporated successfully into OLEDs.

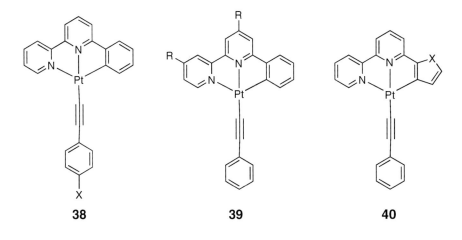

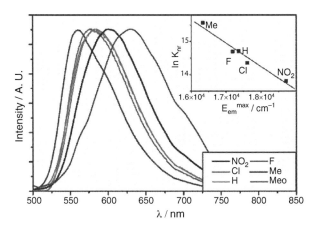

Fig. 18 Normalised emission spectra of complexes of type **38**, carrying the different X groups indicated, in dichloromethane at 298 K. In order of increasing λ_{max}: X = NO$_2$ < Cl < H = F < Me < MeO. *Inset*: plot of ln Σk_{nr} vs the energy corresponding to the respective emission maxima. Reprinted from [74] with permission from ACS

Interestingly, in some cases, a switch to ^{3}LC excited states localised on the acetylide or on the aryl group of the acetylide can be achieved. For example, Fig. 19 shows the influence of polyacetylide ligands: on going from the $n = 3$ to $n = 4$ system **41**, a dramatic switch to a highly structured acetylide-localised state is observed. Similarly, when a 2-pyrenylacetylide ligand is used, the lowest-lying excited state is the ^{3}LC state associated with the pyrenyl moiety, which emits deep into the red (λ_{max} 664 nm).

A somewhat related complex incorporating a pyrenyl unit has been reported recently by Li and Yang, but in this case the pyrenyl unit is substituted onto the central 4-position of the N^N^C ligand, **42** [75]. In this compound, dual emission is

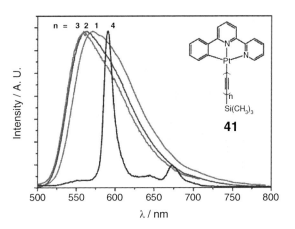

Fig. 19 The influence of extending the polyacetylide ligand in the generic structure shown, revealing the switch from ³MLCT to ³LC emission that occurs at $n = 4$. Reprinted from [74] with permission from ACS

observed from both ³MLCT ($d_{Pt} \rightarrow \pi^*_{NNC}$) and pyrenyl-based $^3\pi$–π^* states, both at 77 K and – unusually – at room temperature. The grow-in and decay kinetics of the two bands, in conjunction with transient absorption spectroscopy, reveal that the pyrenyl $^3\pi$–π^* state is largely populated directly by intersystem crossing from the ¹MLCT state, as is the ³MLCT state, whereas the triplet–triplet internal conversion of the ³MLCT state to the pyrenyl $^3\pi$–π^* state is of negligible importance.

Fillaut et al. have also observed emission from a 'remote' π–π^* state in the complex **43**, where long-lived emission ($\tau \sim 18$ μs) centred around 570 nm is associated with the flavonol unit, whose triplet emission is promoted by coupling to the platinum centre [76]. Interestingly, when the flavonol binds divalent metal

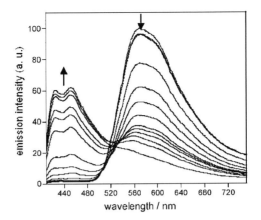

Fig. 20 Luminescence spectral changes of **43** (10^{-5} M) in MeCN upon addition of Pb(ClO$_4$)$_2$ (λ_{ex} = 393 nm) [76]

ions, particularly Pb^{2+}, the relative energies of the excited states change such that the two parts of the molecule become decoupled, and emission then switches to being flavonol-based fluorescence (Fig. 20).

McMillin and co-workers have recently reported on the platinum(II) complex of an N^N^C-coordinating ligand that incorporates a phenazine in the backbone, **44**, together with its cationic N^N^N analogue **45** [77]. They investigated the influence of Lewis acids and bases on the emission. The emission of both complexes is quenched by Lewis bases, as frequently observed for simple Pt(II) complexes, and rationalised on the grounds of sterically accessible attack at the coordinatively unsaturated Pt centre. However, the emission of **44**, but not of **45**, is also quenched by acid. This difference is attributed to the greater degree of charge-transfer character in the cyclometallated system, which is also manifest in its substantially shorter excited state lifetime (τ = 270 ns in chloronaphthalene compared to 5 µs for **45** in dichloromethane).

44 X = C, n = 0
45 X = N, n = 1

46

7.2 Platinum(II) Complexes with N^N^O-Binding Ligands

As noted in Sect. 4.3, the combination of diamine or diimine and phenolic donors in salen and related ligands can lead to emissive complexes. In the same way, the aryl ring of N^N^C-based complexes can be replaced by an *ortho*-phenolate. Thus, a range of luminescent N^N^O-coordinated complexes based on the 6-(2-hydroxyphenyl)-2,2'-bipyridine, have been synthesised by Che and co-workers, **46** (where Y = H, Me or t-Bu; Z = H or F) [78]. In fact, the parent ligand was initially isolated as a side product of the synthesis of the bpy-based analogue of the tetradentate ligand in complex **12**, but the terdentate ligand also proved successful at forming thermally stable, emissive complexes. Broad orange–red emission at room temperature again originates from a mixed $d_{Pt}/p_O \rightarrow \pi^*_{N^{\wedge}N}$ excited state. When incorporated into OLEDs, good performance yellow devices were obtained. Although dopant concentration did not affect the device colour, the presence of fluorine and *tert*-butyl at positions Z and Y respectively was found to give the highest efficiency.

7.3 Platinum(II) Complexes of N^C^N-Binding Ligands

Our group has been investigating complexes of 1,3-di(2-pyridyl)benzene (dpybH) and derivatives. The parent system **47** (Y = Z = H) is the isomer of the N^N^C-coordinated complex (Sect. 7.1). However, the positioning of the cyclometallating aryl ring in the central rather than lateral position within the terdentate system leads to improved luminescence efficiency (e.g. $\phi = 0.60$ for the parent compound), probably due to the increased ligand field strength associated with the shorter Pt–C bond [79]. The highly structured emission that is scarcely shifted in position nor elongated in lifetime upon cooling to 77 K is suggestive of a primarily ligand-centred π–π* state, in which the contribution of the metal is nevertheless sufficient to promote the triplet emission. These conclusions are supported by DFT calculations [10], which reveal a HOMO that spans the cyclometallating ring, the metal and the chloride co-ligand, and a LUMO that is localised on the pyridine rings.

47   **48**

This localisation of frontier orbitals also accounts for the influence of substituents on emission energy. For example, the introduction of aryl substituents into the central 4-position (Y) of the central ring leads to a lowering in the excited state energy, both the singlet state (E_{abs}) and triplet (E_{em}). A good linear correlation of both E_{abs} and E_{em} is observed with the oxidation potential, which can be understood in terms of the influence of increasingly electron-rich pendants in raising the HOMO level, whilst the LUMO is largely unaffected (Fig. 21) [80]. In the case of the most electron-rich amine substituents (e.g. R = –C_6H_4–NMe_2), however, the emissive excited state is stabilised more than anticipated on the basis of the oxidation potential, an observation that has been attributed to a switch to an ICT-type state, involving a high degree of charge transfer, and manifest through the pronounced solvatochromism observed in emission [80, 81]. Reversible switching between the localised and charge-transfer states can be induced by protonation, and the effect can be extrapolated to azacrown pendants that bind divalent metal ions [81].

Whilst the above strategy of introducing electron-rich pendants successfully lowers the excited state energies, scope for shifting to the blue through modification at this position is more limited; an electron-withdrawing ester group leads to a shift of 10 nm. However, further blue shifts have been accomplished by introducing electron-donating substituents into the pyridyl rings, particularly the 4-position, which raise the LUMO energy [82]. Another way of achieving the same effect is to change from pyridine to pyrazole rings, which are poorer π-acceptors. Thus, the emission of complex **48** is blue-shifted by 1,700 cm^{-1} compared to that of **47**,

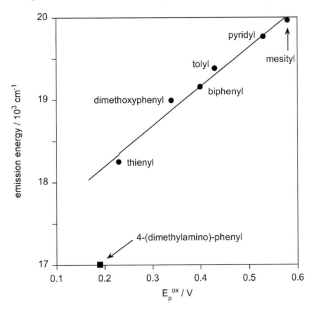

Fig. 21 Correlation of the emission energy E_{em} of complexes of type **47** (Z = H), carrying the aryl substituents indicated at position Y, with the oxidation peak potential measured by cyclic voltammetry in CH_2Cl_2 [80]

though unfortunately accompanied by a substantial decrease in the emission quantum yield [83].

As for the ppy-based systems discussed in Sect. 6.2, the majority of the N^C^N-coordinated complexes display very efficient excimer formation, and the excimers themselves are remarkably highly emissive ($\Phi \sim 0.3$). By varying the dopant concentration within the host emissive layer, the ratio of monomeric (blue) to excimeric (red) emission is varied, and the observed light that the device produces is the net combination. At a doping level of 15% in a CBP host, a device has been obtained that emits with CIE coordinates of 0.337, 0.384, close to white light, and with a quantum efficiency of 18.1%, making it one of the most efficient WOLEDs reported hitherto [84].

In fact, the camel-shaped spectrum leads to a trough of low emission intensity between the monomer and excimer regions, making the coverage across the visible less uniform than might be desirable. Kalinowski et al. have found that this problem can be overcome by 'filling in' this region with emission from an exciplex [85, 86]. The concept is illustrated in Fig. 22. In the emitting region, a starburst amine hole-transporting bathophenanthroline (m-MTDATA) acts as an electron donor (D) to the Pt complex **47** (Y = –CO$_2$Me) as an electron acceptor (A), mixed in a 1:1 ratio. Monomer phosphor triplets, ^{3}A*, their corresponding excimers ^{3}(AA)* formed upon encounter with ground state A, and excited heterodimers, i.e. exciplexes, ^{3}(DA)*, are generated throughout the emissive layer. A colour rendering index of 90 with an external quantum efficiency of 6.5% photons/electron has been attained in this way [85]. The phenomenon of electrophotoluminescence in organics has also been demonstrated for the first time recently using **47** (R = –CO$_2$Me) [87].

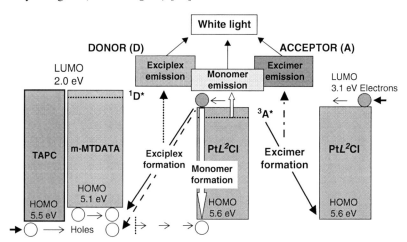

Fig. 22 Generation mechanisms of white light in an OLED based on a hole transporting material (m-MTDATA) acting as an electron donor (D) to an electron acceptor (A) molecule of an organic phosphor (**47** R = CO$_2$Me) mixed in an emissive layer, D:A. The monomer phosphor triplets (^{3}A*), their combination with ground state phosphor acceptor molecules [triplet excimers, ^{3}(AA)*], and excited hetero-dimer [^{3}(DA)*] are generated throughout the emissive layer leading to white light with colour rendering indices (CRI) up to 90 [85]

Luminescent Platinum Compounds: From Molecules to OLEDs   107

These highly emissive complexes have also been applied in sensory applications. Although quenched by O_2, they are not quenched as efficiently as longer-lived systems such as platinum porphyrins. Thus, by combining a thin layer of green-emitting Pt(N^C^N) with a second layer of red-emitting platinum porphyrin in ethyl cellulose, a traffic light response system to O_2 has been developed, in which the system emits red light in the absence of O_2, turning orange and then green as pO_2 increases, and the porphyrin becomes quenched more than the N^C^N complex [88]. Very recently, the simplest complexes of this type have also been shown to be successfully applicable to time-resolved imaging in live cells on the microsecond timescale (Fig. 23) [89]. The parent complex is highly cell permeable, entering essentially under diffusion control, and localises to the nucleoli (Fig. 24). It is also amenable to two-photon excitation in the near-IR.

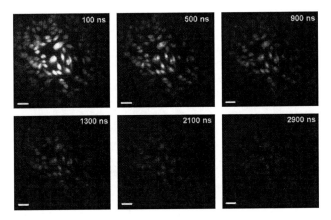

Fig. 23 Time-resolved gated emission images of CHO cells incubated with **47** (Y = Z = H). The images were recorded after 355 nm laser excitation at the time delays shown between 100 and 2,900 ns after the laser flash. The time gate used was 100 ns, exposure time 0.02 s, five accumulations per time delay. *Scale bar*: 50 μm [89]

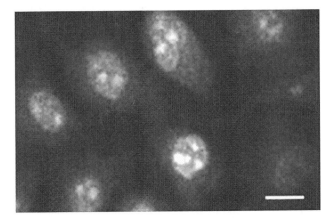

Fig. 24 A two-photon excitation high-resolution emission image of live CHO cells incubated with **47** (Y = Z = H), obtained under 758 nm 180 fs excitation. *Scale bar* = 10 μm [89]

49

Meanwhile, Bruce and co-workers have appended complexes of this type with long alkyl chains, adding interesting liquid crystalline (LC) properties to the luminescence. For example, the properties of an LC state of **49**, obtained by slowly cooling from the isotropic melt to 170°C then rapidly cooling to room temperature, were compared with a non-crystalline phase formed from rapid cooling of the isotropic melt. Structural analysis revealed that the LC phase was composed of a columnar phase of anti-parallel complexes which are independent of one another. The non-crystalline phase, in contrast, consists of a high degree of isotropic grain boundaries, and molecules are therefore in close contact. As a result, emission from the LC state is purely monomeric, and emission from the non-crystalline phase is purely excimeric (Fig. 25). In some cases, reversible thermally-induced switching of excimer to monomer in spin-coated thin films has also been observed [90].

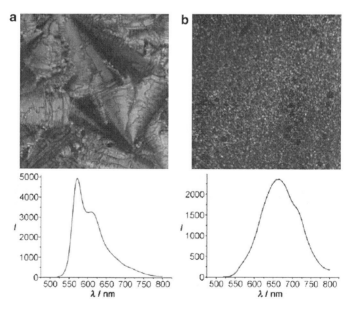

Fig. 25 Photomicrographs taken between cross polarisers (*top*) and emission spectra (*bottom*) of **49** (**a**) fast cooled from the LC phase after the texture is developed, and (**b**) fast cooled from the isotropic phase. Reprinted from [90] with permission from Wiley

References

1. Lakowicz JR (2006) Principles of fluorescence spectroscopy, 3rd edn, chap 1. Springer, Berlin Heidelberg New York
2. Montalti M, Credi A, Prodi L, Gandolfi MT (2006) Handbook of photochemistry, 3rd edn. Taylor & Francis, Boca Raton
3. Hung LS, Chen CH (2002) Mater Sci Eng R 39:143–222
4. Wilson JS, Köhler A, Friend RH, Al-Suti MK, Al-Mandhary MRA, Khan MS, Raithby PR (2000) J Chem Phys 113:7627–7634
5. Baldo MA, O'Brien DF, You Y, Shoustikov A, Sibley S, Thompson ME, Forrest SR (1998) Nature 395:151–154
6. Thompson ME, Burrows PE, Forrest SR (1999) Curr Opin Solid State Mater Sci 4:369–372
7. Danilov EO, Pomestchenko IE, Kinayyigit S, Gentili PL, Hissler M, Ziessel R, Castellano FN (2005) J Phys Chem A 109:2465–2471
8. Michalec JF, Bejune SA, McMillin DR (2000) Inorg Chem 39:2708–2709
9. Hu J, Lin R, Yip JHK, Wong K-Y, Ma D-L, Vittal JJ (2007) Organometallics 26:6533–6543
10. Sotoyama W, Satoh T, Sato H, Matsuura A, Sawatari N (2005) J Phys Chem A 109:9760–9766
11. Williams JAG (2007) Top Curr Chem 281:205–268
12. Miskowski VM, Houlding VH (1989) Inorg Chem 28:1529–1533
13. Miskowski VM, Houlding VH, Che CM, Wang Y (1993) Inorg Chem 32:2518–2524
14. Eastwood D, Gouterman M (1970) J Mol Spectrosc 35:359
15. Castellano FN, Pomestchenko IE, Shikhova E, Hua F, Muro ML, Rajapakse N (2006) Coord Chem Rev 250:1819–1828
16. Kvam P-I, Puzyk MV, Balashev KP, Songstad J (1995) Acta Chem Scand 49:335–343
17. Williams JAG, Develay S, Rochester DL, Murphy L (2008) Coord Chem Rev 252:2596–2611
18. Che C-M, Wan K-T, He L-Y, Poon C-K, Yam VW-W (1989) J Chem Soc Chem Commun 943–944
19. Kunkely H, Vogler A (1990) J Am Chem Soc 112:5625–5627
20. Wan K-T, Che C-M, Cho K-C (1991) J Chem Soc Dalton Trans 1077–1080
21. DeSchryver FC, Collart P, Vandendriessche J, Goedeweeck R, Swinnen A, Van der Auweraer M (1987) Acc Chem Res 20:159–166
22. Kato M, Shishido Y, Ishida Y, Kishi S (2008) Chem Lett 37:16–17
23. Kunkely H, Vogler A (2006) Inorg Chem Commun 9:827–829
24. Keller HJ (ed) (1977) Chemistry and physics of one-dimensional metals. Plenum, New York
25. Miller JS (ed) (1982) Extended linear chain compounds. Plenum, New York
26. Exstrom CL, Sowa JR Jr, Daws CA, Janzen D, Mann KR (1995) Chem Mater 7:15–17
27. Buss CE, Mann KR (2002) J Am Chem Soc 124:1031–1039
28. Kato M (2007) Bull Chem Soc Jpn 80:287–294
29. Sun Y, Ye K, Zhang H, Zhang J, Zhao L, Li B, Yang G, Yang B, Wang Y, Lai S-W, Che C-M (2006) Angew Chem Int Ed 45:5610–5613
30. Zhou X, Zhang H-X, Pan Q-J, Li M-X, Wang Y, Che C-M (2007) Eur J Inorg Chem 2181–2188
31. Klein A, Kaim W (1995) Organometallics 14:1176–1186
32. Dungey KE, Thompson BD, Kane-Maguire NAP, Wright LL (2000) Inorg Chem 39:5192–5196
33. Nishida J, Maruyama A, Iwata T, Yamashita Y (2005) Chem Lett 34:592–593
34. De Crisci AG, Lough AJ, Multani K, Fekl U (2008) Organometallics 27:1765–1779
35. Ardasheva LP, Shagisultanova GA (1998) Russ J Inorg Chem 43:85–93
36. Che C-M, Chan S-C, Xiang H-F, Chan MCW, Liu Y, Wang Y (2004) Chem Commun 1484–1485
37. Lin Y-Y, Chan S-C, Chan MCW, Hou Y-J, Zhu N, Che C-M, Liu Y, Wang Y (2003) Chem Eur J 9:1263–1272
38. Lü X, Wong W-Y, Wong W-K (2008) Eur J Inorg Chem 523–528

39. Chai W-L, Jin W-J, Lü X-Q, Bi W-Y, Song J-R, Wong W-K, Bao F (2008) Inorg Chem Commun 11:699–702
40. Chang S-Y, Kavitha J, Li S-W, Hsu C-S, Chi Y, Yeh Y-S, Chou P-T, Lee G-H, Carty AJ, Tao Y-T, Chien C-H (2006) Inorg Chem 45:137–146
41. Chang S-Y, Kavitha J, Hung J-Y, Chi Y, Cheng Y-M, Li E-Y, Chou P-T, Lee G-H, Carty AJ (2007) Inorg Chem 46:7064–7074
42. Xiang H-F, Chan S-C, Wu KK-Y, Che C-M, Lai PT (2005) Chem Commun 1408–1409
43. Umakoshi K, Kojima T, Saito K, Akatsu S, Onishi M, Ishizaka S, Kitamura N, Nakao Y, Sakaki S, Ozawa Y (2008) Inorg Chem 47:5033–5035
44. Chassot L, Müller E, von Zelewsky A (1984) Inorg Chem 23:4249–4253
45. Maestri M, Sandrini D, Balzani V, Chassot L, Jolliet P, von Zelewsky A (1985) Chem Phys Lett 122:375–379
46. Barigelletti F, Sandrini D, Maestri M, Balzani V, von Zelewsky A, Chassot L, Jolliet P, Maeder U (1988) Inorg Chem 27:3644–3647
47. Cocchi M, Virgili D, Sabatini C, Fattori V, Di Marco P, Maestri M, Kalinowski J (2004) Synth Met 147:253–256
48. Cocchi M, Fattori V, Virgili D, Sabatini C, Di Marco P, Maestri M, Kalinowski J (2004) Appl Phys Lett 84:1052–1054
49. Black DSC, Deacon GB, Edwards GL (1994) Aust J Chem 47:217–227
50. Godbert N, Pugliese T, Aiello I, Bellusci A, Crispini A, Ghedini M (2007) Eur J Inorg Chem:5105–5111
51. Niedermair F, Waich K, Kappaun S, Mayr T, Trimmel G, Mereiter K, Slugovc C (2007) Inorg Chim Acta 360:2767–2777
52. Brooks J, Babayan Y, Lamansky S, Djurovich PI, Tsyba I, Bau R, Thompson ME (2002) Inorg Chem 41:3055–3066
53. Mdleleni MM, Bridgewater JS, Watts RJ, Ford PC (1995) Inorg Chem 34:2334–2442
54. Kovelenov YA, Blake AJ, George MW, Matousek P, Mel'nikov MY, Parker AW, Sun X-Z, Towrie M, Weinstein JA (2005) Dalton Trans:2092–2097
55. Yin B, Niemeyer F, Williams JAG, Jiang J, Boucekkine A, Toupet L, Le Bozec H, Guerchais V (2006) Inorg Chem 45:8584–8596
56. Shavaleev NM, Adams H, Best J, Edge R, Navaratnam S, Weinstein JA (2006) Inorg Chem 45:9410–9415
57. Niedermair F, Kwon O, Zojer K, Kappaun S, Trimmel G, Mereiter K, Slugovc C (2008) Dalton Trans 4006–4014
58. Ballardini R, Indelli MT, Varani G, Bignozzi CA, Scandola F (1978) Inorg Chim Acta 31: L423–L424
59. Yang C-J, Yi C, Xu M, Wang J-H, Liu Y-Z, Gao X-C, Fu J-W (2006) Appl Phys Lett 89:233506
60. D'Andrade BW, Brooks J, Adamovich V, Thompson ME, Forrest SR (2002) Adv Mater 14:1032–1036
61. U.S. Department of Energy (2003) Illuminating the challenges: solid state lighting program planning workshop report. US Government Printing Office, Washington DC
62. Kido J, Shionoya H, Nagai K (1995) Appl Phys Lett 67:2281–2283
63. Kido J, Kimura M, Nagai K (1995) Science 267:1332–1334
64. Adamovich V, Brooks J, Tamayo A, Alexander AM, Djurovich PI, D'Andrade BW, Adachi C, Forrest SR, Thompson ME (2002) New J Chem 26:1171–1178
65. Wong W-Y, He Z, So S-K, Tong K-L, Li Z (2005) Organometallics 24:4079–4082
66. He Z, Wong W-Y, Yu X, Kwok H-S, Lin Z (2006) Inorg Chem 45:10922–10937
67. Zhou G-J, Wang X-Z, Wong W-Y, Yu X-M, Kwok H-S, Lin Z (2007) J Organomet Chem 692:3461–3473
68. Cho J-Y, Domercq B, Barlow S, Suponitsky KY, Li J, Timofeeva TV, Jones SC, Hayden LE, Kimyonok A, South CR, Weck M, Kippelen B, Marder SR (2007) Organometallics 26:4816–4829

69. Balzani V, Carassiti V (1968) J Phys Chem 72:383–388
70. Andrews LJ (1979) J Phys Chem 83:3203–3209
71. McMillin DR, Moore JJ (2002) Coord Chem Rev 229:113–121
72. Wong KM-C, Yam VW-W (2007) Coord Chem Rev 251:2477–2488
73. Lai S-W, Chan MC-W, Cheung T-C, Peng S-M, Che C-M (1999) Inorg Chem 38:4046–4055
74. Lu W, Mi B-X, Chan MCW, Hui Z, Che C-M, Zhu N, Lee S-T (2004) J Am Chem Soc 126:4958–4971
75. Wang A, Xiong F, Morlet-Savary F, Li S, Li Y, Fouassier J-P, Yang G (2008) J Photochem Photobiol A 194:230–237
76. Lanoë P-H, Fillaut J-L, Toupet L, Williams JAG, Le Bozec H, Guerchais V (2008) Chem Commun 4333–4335
77. McGuire R Jr, Wilson MH, Nash JJ, Fanwick PE, McMillin DR (2008) Inorg Chem 47:2946–2948
78. Kwok C-C, Ngai HMY, Chan S-C, Sham IHT, Che C-M, Zhu N (2005) Inorg Chem 44:4442–4444
79. Williams JAG, Beeby A, Davies ES, Weinstein JA, Wilson C (2003) Inorg Chem 42:8609–8611
80. Farley SJ, Rochester DL, Thompson AL, Howard JAK, Williams JAG (2005) Inorg Chem 44:9690–9703
81. Rochester DL, Develay S, Zalis S, Williams JAG (2009) Dalton Trans: 1728–1741
82. Fattori V, Williams JAG, Murphy L, Cocchi M, Kalinowski J (2008) Photonics Nanostruct Fundam Appl 6:225–230
83. Develay S, Blackburn O, Thompson AL, Williams JAG (2008) Inorg Chem 47:11129–11142
84. Cocchi M, Kalinowski J, Virgili D, Fattori V, Develay S, Williams JAG (2007) Appl Phys Lett 90:023506
85. Kalinowski J, Cocchi M, Virgili D, Fattori V, Williams JAG (2007) Adv Mater 19:4000–4005
86. Virgili D, Cocchi M, Fattori V, Sabatini C, Kalinowski J, Williams JAG (2006) Chem Phys Lett 433:145–149
87. Kalinowski J, Cocchi M, Virgili D, Fattori V, Williams JAG (2007) Chem Phys Lett 447:279–283
88. Evans RC, Douglas P, Williams JAG, Rochester DL (2006) J Fluorescence 16:201–206
89. Botchway SW, Charnley M, Haycock JW, Parker AW, Rochester DL, Weinstein JA, Williams JAG (2008) Proc Natl Acad Sci U S A 105:16071–16076
90. Kozhevnikov VN, Donnio B, Bruce DW (2008) Angew Chem Int Ed 47:1–5

Luminescent Iridium Complexes and Their Applications

Zhiwei Liu, Zuqiang Bian, and Chunhui Huang

Abstract Considerable studies have been made on iridium complexes during the past 10 years, due to their high quantum efficiency, color tenability, and potential applications in various areas. In this chapter, we review the synthesis, structure, and photophysical properties of luminescent Ir complexes, as well as their applications in organic light-emitting diodes (OLEDs), biological labeling, sensitizers of luminescence, and chemosensors.

Keywords Biological labeling, Chemosensor, Iridium complex, Luminescent, Organic light-emitting diodes, Structure, Synthesis

Contents

1 Introduction 114
2 Synthesis and Structure 114
   2.1 Neutral Iridium Complex 114
   2.2 Ionic Iridium Complex 118
3 Photophysical Properties 120
   3.1 Neutral Iridium Complex 120
   3.2 Ionic Iridium Complex 123
4 Applications of Luminescent Iridium Complexes 126
   4.1 Applications in OLEDs 126
   4.2 Biological Labeling Reagents 131
   4.3 Sensitizer of Lanthanide Luminescence 133
   4.4 Sensor Applications 135
5 Summaries and Outlook 138
References 138

Z. Liu, Z. Bian, and C. Huang (✉)
Beijing National Laboratory for Molecular Sciences (BNLMS), State Key Laboratory of Rare Earth Materials Chemistry and Applications, Peking University, Beijing 100871, People's Republic of China
e-mail: chhuang@pku.edu.cn

1 Introduction

There has been growing interest in luminescent iridium (Ir) complexes due to their high quantum efficiency and color tunability. The highly efficient emission is attributed to strong spin-orbit coupling caused by the presence of the 5d metal ion, which leads to efficient intersystem crossing from a singlet to a triplet excited state. Mixing of the singlet and triplet excited states via spin-orbit coupling relaxes the spin-forbidden nature of the radiative relaxation of the triplet state [1, 2]. Color diversity arises because excited states in Ir complexes are ligand-related. The energy of the lowest excited state can therefore be controlled by deliberately adjusting the energy of ligand orbitals through substituent effects [3, 4] or by entirely changing the cyclometalated ligand structure [5, 6].

Various cyclometalated Ir complexes have been reported, which can be classified into two main groups. The first group is neutral, containing Ir(C^N)$_3$ [2, 4, 7–18] (C is a cyclometalated carbon and N is a heterocyclic nitrogen) and Ir(C^N)$_2$(LX) [6, 19–28] (LX represents an ancillary ligand) type with tris-bidentate ligands, and Ir(N^C^N)(C^N^X) (X is an anionic ligand or cyclometalated carbon) [29–31] type with bi–tridentate ligands. The other group is ionic, including [Ir(C^N)$_2$(N^N)]$^+$ [32–40] type with tris-bidentate ligands, [Ir(C^N)$_2$(L)$_2$]$^-$ [41] (L denotes an anionic ancillary ligand) type with unidentate and bidentate ligands, and [IrN$_{6-n}$C$_n$]$^{(3-n)+}$ ($n = 0, 1, 2$) [31, 42–44] type with tridentate ligands. Because of their rich photophysical properties, Ir complexes have been widely studied in many applications, such as organic light-emitting diodes (OLEDs) [1, 2, 6, 8, 10, 13, 19, 20, 24, 26, 41, 45–51], light-emitting electrochemical cells (LECs) [52–60], biological labeling [33–36, 61–64], sensitizers of luminescence [65–67], and chemosensors [11, 68–71].

The synthesis, structure, and photophysical properties of luminescent Ir complexes, as well as their applications are introduced in this chapter.

2 Synthesis and Structure

2.1 Neutral Iridium Complex

2.1.1 Ir(C^N)$_3$ Type

In 1985, Watts et al. reported the first Ir(C^N)$_3$-type Ir complex *facial*-Ir(ppy)$_3$ (*fac*-**1**), which was formed in 10% yield as a side product in the reaction of Hppy (2-phenylpyridine) with hydrated IrCl$_3$ [16]. They subsequently attempted to extend this procedure with methyl-substituted ppy ligands, but only trace amounts of the desired complex were obtained [4]. In 1991, they developed a new procedure for the high-yield synthesis of *fac*-Ir(C^N)$_3$ with ppy and other substituted ppy ligands [4]. The procedure utilized the starting material Ir(acac)$_3$ (acac =

2,4-pentanedionate) instead of hydrated $IrCl_3$ (Fig. 1). The method typically produced *fac*-Ir(C^N)$_3$ in high yields of 40–75%, but did not prepare complexes for other structural cyclometalated ligands. To solve this problem, Güdel et al. reported another method in 1994 which involved treating a μ-dichloro bridged dimer complex [Ir(C^N)$_2$Cl]$_2$ with excess HC^N (free cyclometalated ligand). They found that *fac*-Ir(C^N)$_3$ complexes containing different cyclometalated ligands could be synthesized in high yield [17].

Since the initial report in 1999 using *fac*-**1** as an emitter achieved highly efficient phosphorescent OLEDs, many Ir(C^N)$_3$-type Ir complexes, including small molecular complexes [8, 9, 15] and dendritic complexes [72–78], have been reported. The synthesis and structure of these complexes have been thoroughly investigated.

In general, there are three synthetic methods to prepare Ir(C^N)$_3$-type complexes (Fig. 2). Methods B and C have several advantages over method A. For example, Ir(C^N)$_2$(O^O) (O^O = 2,2,6,6-tetramethyl-3,5-heptanedionate) and [Ir(C^N)$_2$Cl]$_2$ compounds are easily prepared in high yield from a less expensive starting material; methods B and C give higher yields than method A in the last reaction step [15].

Meridional (abbreviated as *mer*) isomers of Ir(C^N)$_3$ complexes can also be prepared as a different steric geometry configuration of the facial isomers (Fig. 3). Results show that *fac*-Ir(C^N)$_3$ tends to form under higher temperature (suggesting that *fac* isomers are thermodynamically favored products), whereas *mer*-Ir(C^N)$_3$ can be obtained at lower temperature (suggesting that *mer* isomers are kinetically favored products). This is consistent with the experimental phenomenon that the *mer* isomer can be converted to the *fac* isomer under treatment at high temperature [15].

Meridional and *fac* isomers are different not only in steric geometry configuration, but also in Ir–N and Ir–C bond lengths. Key bond lengths of *fac*-Ir(tpy)$_3$ (*fac*-**2**,

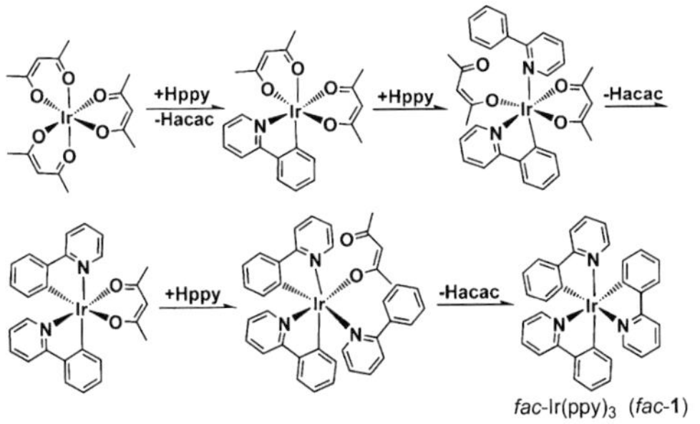

Fig. 1 Schematic representation of the reaction mechanism in the synthesis of Ir(C^N)$_3$ type complex *fac*-**1**

$$\mathrm{Ir(acac)_3 + 3\ HC{^\wedge}N \xrightarrow{\text{glycerol}} Ir(C{^\wedge}N)_3} \quad \text{A}$$

$$\mathrm{Ir(C{^\wedge}N)_2(O{^\wedge}O) + HC{^\wedge}N \xrightarrow{\text{glycerol}} Ir(C{^\wedge}N)_3} \quad \text{B}$$

$$\mathrm{(C{^\wedge}N)_2Ir(u\text{-}Cl)_2(C{^\wedge}N)_2 + 2\ HC{^\wedge}N \xrightarrow{\text{glycerol, } K_2CO_3} Ir(C{^\wedge}N)_3} \quad \text{C}$$

Fig. 2 Three synthetic methods to prepare Ir(C^N)$_3$ type Ir complex

Fig. 3 Steric configuration for *fac*-Ir(C^N)₃ and *mer*-Ir(C^N)₃

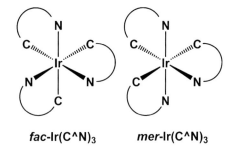

fac- Ir(tpy)₃ (*fac*-2)

Bond type	Bond length (Å)	
	fac-2	*mer*-2
Ir-N1		2.151(9)
Ir-N2	2.132(5)	2.044(8)
Ir-N3		2.065(8)
Ir-C1		2.076(10)
Ir-C2	2.024(5)	2.086(12)
Ir-C3		2.020(8)

mer- Ir(tpy)₃ (*mer*-2)

Fig. 4 Crystal structures and key bond lengths of Ir(C^N)₃ type Ir complexes *fac*-2 and *mer*-2

tpy is 4-methylphenylpyridine) and *mer*-Ir(tpy)₃ (*mer*-2) are compared in Fig. 4. In the *fac* isomer, all the Ir–C bonds are *trans* to a pyridyl group, and the Ir–N bonds are *trans* to a phenyl group, leading to identical Ir–C and Ir–N bond lengths. In the *mer* isomer, some of the bond lengths differ markedly from those of the *fac* isomer. This may be because Ir–C (or Ir–N) bonds in the *mer* isomer share an identical/different electronic environment to Ir–C (or Ir–N) bonds in the *fac* isomer.

2.1.2 Ir(C^N)₂(LX) Type

The universal synthetic method used to prepare an Ir(C^N)₂(LX)-type complex is shown in Fig. 5. Dimer complex [Ir(C^N)₂Cl]₂ is readily prepared from reaction of the ligand precursor and IrCl₃·nH₂O [79], and chloro-ion ligands can be subsequently replaced with an LX chelate. The most studied ancillary ligand LX is acac, but it can be varied with other monoanionic bidentate ligands, such as picolinic acid, *N*-methylsalicylimine [23], 2-(5-phenyl-4*H*-[1,2,4]triazol-3-yl)-pyridine [28], 2,2,6,6-tetramethyl-3,5-heptanedionate, 1-phenyl-4,4-dimethyl-1,3-pentanedionate, 1,3-diphenyl-1,3-propanedionate, pyrazolyl, pyrazolyl-borate [80], and (2-pyridyl) pyrazolate derivatives [81].

To understand the structure of the Ir(C^N)₂(LX)-type Ir complex, the crystal structure of Ir(tpy)₂(acac) (**3**, Fig. 6) was compared with *fac*-2 and *mer*-2 mentioned above. The bis-cyclometalated fragment of **3** has the same disposition of tpy ligands

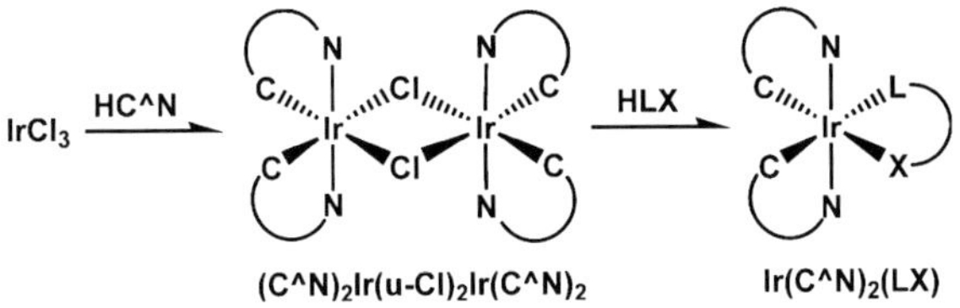

Luminescent Iridium Complexes and Their Applications

Fig. 5 Synthetic route for Ir(C^N)₂(LX) type complex

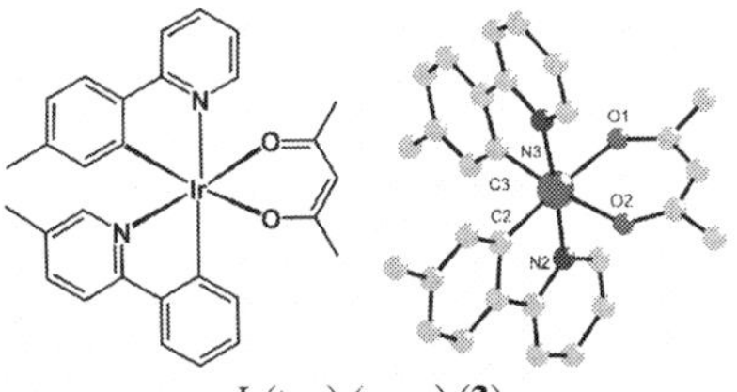

Fig. 6 Chemical and crystal structures of Ir(C^N)₂(LX) type complex **3**

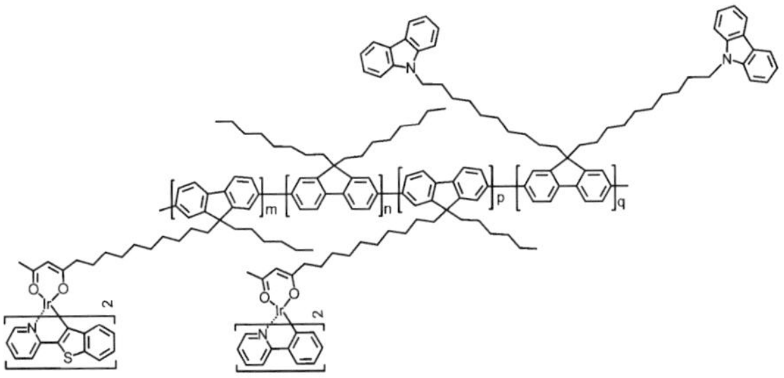

Fig. 7 Chemical structure of a polymerized Ir(C^N)₂(LX)-type Ir complex

as found in *mer*-**2**, and the mutually *trans*-disposed Ir–N bonds in both complexes have similar lengths (av = 2.032(5) Å). The weak *trans* influence of the acac ligand leads to shorter Ir–C bonds (av = 1.984(6) Å) for the complex **3** than those observed in complex *fac*-**2** or *mer*-**2**.

There is also a polymerized Ir(C^N)₂(LX)-type Ir complex in which the bis-cyclometalated fragment Ir(C^N)₂ is incorporated on the side chain of a polymer. The design of the entire system could have better dissolution and film-forming ability as that of the polymer, as well as high quantum efficiency similar to that of luminescent Ir complexes. The chemical structure of a selected polymerized Ir(C^N)₂(LX)-type Ir complex is shown in Fig. 7 [82].

2.1.3 Ir(N^C^N)(C^N^X) Type

Neutral Ir complexes mentioned above are all composed of bidentate ligands. Believing tridentate ligands can lead to different excited state, stereoisomer, and

Fig. 8 Synthetic methods of Ir(N^C^N)(C^N^X) type complexes **4** and **5**

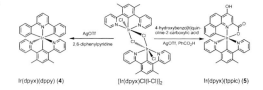

linear assembly, some groups focused their research on Ir complex with tridentate ligands. Figure 8 shows the synthetic methods of Ir(dpyx)(dppy) (**4**) and Ir(dpyx)(tppic) (**5**), two neutral Ir complexes with tridentate ligands. Complex **4** was obtained by heating the chloro-bridged dimer [Ir(dpyx)Cl(l-Cl)]$_2$ with AgOTf in molten 2,6-diphenylpyridine (dppyH), followed by rapid chromatographic purification [30], while complex **5** was synthesized upon reaction of the dimer with 4-hydroxybenzo[h]quinoline-2-carboxylic acid (tppicH) using molten benzoic acid as the solvent [29]. The N^C^N tridentate coordination mode of dpyx was confirmed in [Ir(dpyx)-(DMSO)Cl$_2$] (DMSO is dimethyl sulfoxide) by X-ray crystal structure [29].

2.2 Ionic Iridium Complex

2.2.1 [Ir(C^N)$_2$(N^N)]$^+$ Type

The cyclometalated Ir complexes investigated earlier with a formula of [Ir(C^N)$_2$(N^N)]$^+$ are complexes [Ir(ppy)$_2$(bpy)]Cl (**6**) and [Ir(ppy)$_2$(bpy)]PF$_6$ (**7**) (bpy and PF$_6$ represent 2,2′-bipyridine and hexafluorophosphate, respectively) [39]. The two cationic Ir complexes were prepared by a modified method employed by Nonoyama [83]. The detailed synthetic route is shown in Fig. 9. The stereo structure of these [Ir(C^N)$_2$(N^N)]$^+$ complexes is similar to those of Ir(C^N)$_2$(LX) complexes, with a *trans-N,N* configuration of the C^N ligands, whereas the diimine ligand is located opposite *cis*-oriented carbon atoms, completing an octahedral arrangement [84].

2.2.2 [Ir(C^N)$_2$(L)$_2$]$^−$ Type

An [Ir(C^N)$_2$(L)$_2$]$^−$ type anionic Ir complex can be conveniently synthesized in low-boiling solvent by reacting the corresponding [Ir(C^N)$_2$Cl]$_2$ dimer complex with a pseudohalogen ligand such as tetrabutylammonium cyanide, tetrabutylammonium thiocyanate, or tetrabutylammonium cyanate [41, 85]. Figure 10 shows the chemical and crystal structure of the complex (C$_4$H$_9$)$_4$N[Ir(ppy)$_2$(CN)$_2$] (**8**) [41, 85]. The Ir atom is octahedrally coordinated by four ligands, with the N atoms of the

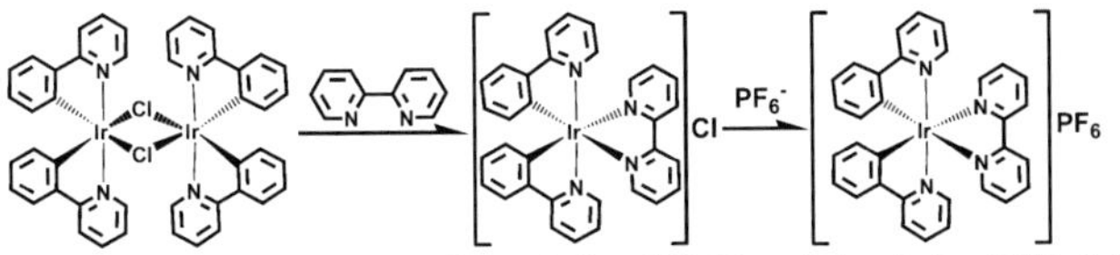

$Ir_2(ppy)_4Cl_2$ $[Ir(ppy)_2(bpy)]Cl$ (**6**) $[Ir(ppy)_2(bpy)]PF_6$ (**7**)

Fig. 9 Synthetic route to $[Ir(C\text{^}N)_2(N\text{^}N)]^+$ type complexes **6** and **7**

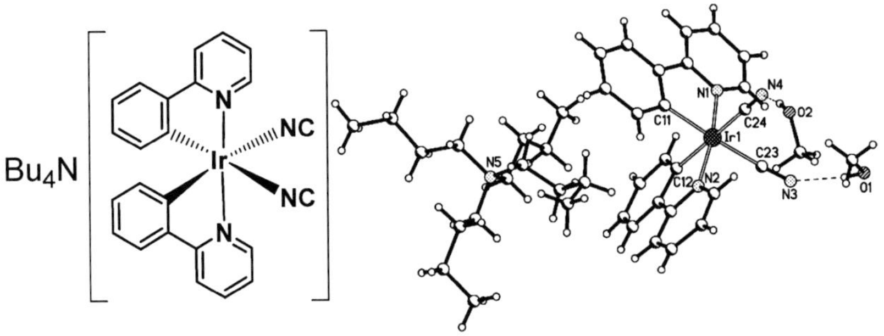

$(C_4H_9)_4N[Ir(ppy)_2(CN)_2]$ (**8**)

Fig. 10 Chemical and crystal structures of $[Ir(C\text{^}N)_2(L)_2]^-$ type complex **8**

2-phenylpyridine ligands in a *trans* disposition similar to that in $[Ir(ppy)_2Cl]_2$, whereas cyanide ligands coordinate through the carbon atom and adopt a *cis* configuration.

These complexes are stable at room temperature as a solid and in solution containing noncoordinating solvents such as dichloromethane, chloroform, methanol, or ethanol. This type of Ir complex may undergo slow substitution of the pseudohalogen ligands upon standing for days in a strong coordinating solvent such as dimethyl sulfoxide [41].

2.2.3 $[IrN_{6-n}C_n]^{(3-n)+}$ ($n = 0, 1, 2$) Type

Figure 11 shows chemical structures for $[IrN_{6-n}C_n]^{(3-n)+}$ ($n = 0, 1, 2$) type Ir complexes with tridentate ligands. Complex **9** was first reported by Degraff and coworkers, with fusion reaction and arduous purification [86]. Subsequently, a milder and stepwise route was described by Sauvage, involving initial reaction of terpyridine (tpyH) with $IrCl_3 \cdot nH_2O$ in ethylene glycol to give $[Ir(tpy)Cl_3]$ as an intermediate, followed by reaction with a second equivalent of tpy in refluxing ethylene glycol [87]. Purification of the tricationic complex is normally best achieved after ion exchange to the hexafluorophosphate salt, which offers the advantage of being soluble in polar organic solvents for column chromatography [31]. In order to obtain complexes **10** and **11**, compound 1,3-di(2-pyridyl)-4,6-dimethylbenzene (dpyxH) was employed to react with $IrCl_3 \cdot nH_2O$ to give a chloro-bridged dimer $[Ir(dpyx)Cl(\mu\text{-}Cl)]_2$, reactions of the dimer with 4′-tolylterpyridine (tppyH) and 6-phenyl-2,2′-bipyridine (phbpyH)

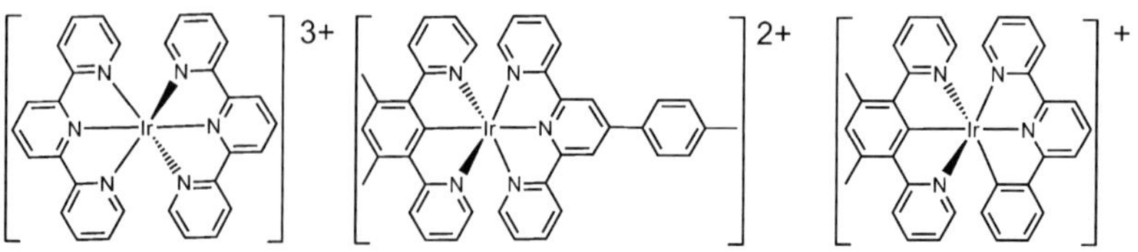

[Ir(tpy)₂]³⁺ (**9**)   [Ir(dpyx)(tpy)]²⁺ (**10**)   [Ir(dpyx)(phbpy)]⁺ (**11**)

Fig. 11 Chemical structures of $[IrN_{6-n}C_n]^{(3-n)+}$ ($n = 0, 1, 2$) type complexes **9–11**

gives complex **10** and **11**, respectively [29, 30]. An X-ray diffraction study of a single crystal of the former confirms the mutually orthogonal orientation of the two ligands, each bound tridentately [30].

3 Photophysical Properties

3.1 Neutral Iridium Complex

3.1.1 Ir(C^N)₃ Type

The photophysical properties of Ir(C^N)₃ have been examined by several research groups [8, 15]. Two principal transitions are observed in this type of Ir complex: (1) metal-to-ligand charge transfer (MLCT) in which an electron is promoted from a metal d orbital to a vacant π^* orbital on one of the ligands and (2) ligand-centered (LC) transitions in which an electron is promoted between π orbitals on one of the coordinated ligands. Phosphorescence in Ir(C^N)₃-type complexes enabled by strong spin-orbit coupling mainly arise from a mixture of ³LC and ³MLCT excited states, whereas the emissive excited state is predominantly the excited state having the lowest energy.

The lowest excited state of a Ir(C^N)₃-type complex can be tuned by changing the substitution of electron-donating or electron-withdrawing groups on the cyclometalated ligand, or by entirely changing the cyclometalated ligand structure [15]. Table 1 lists the photophysical properties of complexes **1** (Fig. 1), **2** (Fig. 4), Ir(46dfppy)₃ (**12**), Ir(ppz)₃ (**13**), Ir(46dfppz)₃ (**14**), and Ir(tmfppz)₃ (**15**) (Fig. 12). Photophysical properties of complex **2** (**14**) are different from that of **12** (**15**), but both are similar to those of parent compound **1** (**13**). This is unsurprising because substitution of donor or acceptor groups tunes the highest occupied molecular orbital (HOMO) and the lowest unoccupied molecular orbital (LUMO) levels of the metal complex in parallel, leading to marginal changes in maximal emission. Photophysical properties of complex **13** are entirely changed compared with complex **1**, which is ascribed to skeletal change of the cyclometalated ligand.

Table 1 Photophysical properties of complexes **1, 2, 12–15** in 2-methyltetrahydrofuran

Complex	Emission at 77 K		Emission at 298 K		
	λ_{max}	τ (µs)	λ_{max}	τ (µs)	Φ_{PL}
fac-**1**	492	3.6	510	1.9	0.40
Mer-**1**	493	4.2	512	0.15	0.036
fac-**2**	492	3.0	510	2.0	0.50
Mer-**2**	530	4.8	550	0.26	0.051
fac-**12**	450	2.5	468	1.6	0.43
Mer-**12**	460	5.4	482	0.21	0.053
fac-**13**	414	14	–	–	–
mer-**13**	427	28	–	–	–
fac-**14**	390	27	–	–	–
mer-**14**	402	33	–	–	–
fac-**15**	422	17	428	0.05	–
mer-**15**	430	32	–	–	–

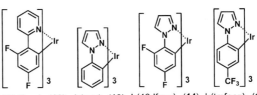

Fig. 12 Chemical structures of complexes **12–15**

Ir(46dfppy)₃ (**12**) Ir(ppz)₃ (**13**) Ir(46dfppz)₃ (**14**) Ir(tmfppz)₃ (**15**)

Photophysical properties of *mer*-Ir(C^N)₃-type complexes are very different to those of *fac*-Ir(C^N)₃-type complexes (Table 1). Meridional ones traditionally exhibit a broad, red-shifted emission, and lower quantum efficiencies, which can be attributed to the reason that the *mer* configuration usually arises from strongly *trans* influencing phenyl groups being opposite each other [88].

3.1.2 Ir(C^N)₂(LX) Type

The lowest triplet energy level of the ancillary ligand LX lie well above the energies of LC and MLCT excited states in most of the Ir(C^N)₂(LX)-type complexes, so luminescence of Ir(C^N)₂(LX) is dominated by ³LC and ³MLCT transitions. This leads to similar phosphorescence emission to the *fac*-Ir(C^N)₃ complexes with the same cyclometalated ligand [6, 23]. In such cases, density functional theory (DFT) calculations indicate that HOMOs are largely metal-centered, whereas LUMOs are primarily localized on the heterocyclic rings of the cyclometalated ligand. The ancillary is therefore not directly involved in the lowest excited state.

Though the ancillary ligand is not directly involved in the lowest energy excited state when it has a higher triplet energy level, it can alter the excited energy state by modifying electron density at the metal center [80]. Figure 13 shows the emission spectra of Ir(tpy)₂(pz₂Bpz₂) (**16**), Ir(tpy)₂(pz₂H) (**17**), Ir(tpy)₂(pzH)Cl (**18**), and Ir(tpy)₂(acac) (**3**) at room temperature. All have the same "Ir(tpy)₂" fragment, but there is a clear blue-shift in the maximum emission wavelength as the ancillary

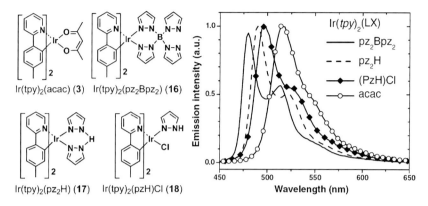

Fig. 13 Chemical structures of **16–18**, and their photoluminescent spectra in 2-methyltetrahydrofuran compared with **3** at room temperature

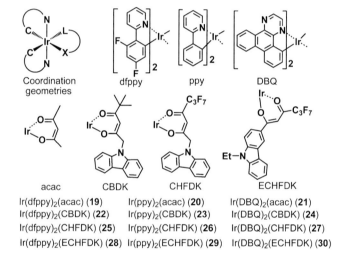

Fig. 14 Chemical structures of complexes **19–30**

ligand is changed. This is attributed to a net electron-withdrawing ancillary ligand that pulls electron density away from the Ir atom, thereby stabilizing the metal orbitals and lowering the HOMO energy; whereas the LUMO of the Ir(tpy)$_2$(LX) complexes, localized on the pyridyl rings, is expected to be largely unaffected with respect to energy. The HOMO-LUMO gap should consequently increase for stronger electron-withdrawing ancillary ligands, leading to a blue-shift in emission.

If the lowest triplet energy level of the LX ligand is lower in energy than the ^{3}LC or ^{3}MLCT, it will be the lowest energy excited state, and thus a switch from "Ir(C^N)$_2$" to LX-based emission can be observed. Our recent work confirms this hypothesis [89]. Twelve Ir complexes (complexes **19–30**, Fig. 14) with general formula Ir(C^N)$_2$(LX) were synthesized by changing the triplet energy level of the

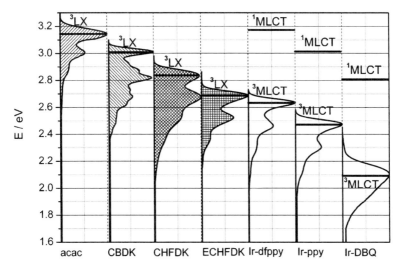

Fig. 15 State density of triplet energy diagram for the β-diketonates and Ir(C^N)$_2$ fragments of complexes **19–30**

cyclometalated and the β-diketonate ligands, and photophysical properties were compared. Strong ^{3}LC- or ^{3}MLCT-based phosphorescence was observed if triplet-state density maps of the β-diketonate and the Ir(C^N)$_2$ fragment were not superimposed (triplet energy of the former is high). If the triplet-state density map of the two parts was superimposed, ^{3}LC- or ^{3}MLCT-based transition would be quenched at room temperature, and DFT calculations show that the lowest excited state is determined by the cyclometalated and ancillary ligand (Fig. 15).

Interligand electron transfer (ILET) from the cyclometalated ligand to the ancillary ligand is also possible in some excited Ir(C^N)$_2$(LX) complexes, particularly if the triplet energy of the ancillary ligand is much lower. Park et al. observed emission across the visible spectrum by changing the structure of the ancillary ligand in a series of complexes undergoing ILET before emission [90–92].

3.2 Ionic Iridium Complex

3.2.1 [Ir(C^N)$_2$(N^N)]$^+$ Type

The photophysical properties of [Ir(C^N)$_2$(N^N)]$^+$-type cationic Ir complexes are complicated. In some excited complexes, the neutral, diimine and cyclometalated ligands provide orbitals that participate in excited-state transitions. The cyclometalated ligand tends to be associated with the ^{3}LC transition, and the diimine ligand with the ^{3}MLCT transition [93]. In such cases, the excited state of the complex can be tuned directly through ligand modification because each ligand is linked to a different transition. Nazeeruddin et al. investigated three cationic Ir complexes **31–33** (Fig. 16) by modulating the electronic structure of the complex using selective ligand functionalization [94]. They developed a strategy to tune the phosphorescence wavelength for this class of compound [94]. The electron-withdrawing substituent on the C^N ligand decreased donation to the metal and therefore stabilized the metal-based HOMO, whereas the electron-releasing substituent on the N^N ligand led to destabilization of the N^N ligand-based LUMO, ultimately leading to increased HOMO-LUMO gaps and emission energies (Table 2).

It was proved that photophysical properties of $[Ir(C^\wedge N)_2(N^\wedge N)]^+$-type complexes can also be adjusted by modifying the conjugated lengths of the diimine ligand [84]. Theoretical calculations, photophysical studies and electrochemical studies of a series of cationic Ir complexes **34–39** (Fig. 17) showed that their excited

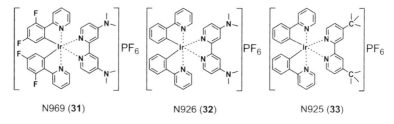

Fig. 16 Chemical structures of **31–33**

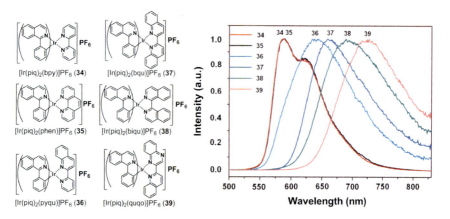

Fig. 17 Chemical structures and emission spectra of **34–39**

Table 2 Photophysical properties of complexes **31–33** in acetonitrile at 298 K

Complex	λ_{max}	Φ_{PL}	τ (μs)
31	463, 493	0.85 ± 0.1	4.11 ± 0.02
32	491, 520	0.80 ± 0.1	2.43 ± 0.02
33	581	0.23	0.557

states simultaneously contain ^{3}MLCT, triplet ligand-to-ligand charge transfer (^{3}LLCT), and ^{3}LC transitions. Their emission wavelengths can therefore be tuned significantly (~150 nm) by changing the conjugated length of N^N ligands.

3.2.2 [Ir(C^N)$_2$(L)$_2$]$^-$ Type

The structures of cyclometalated ligand C^N and spectator ligand L should be considered when evaluating the photophysical properties of [Ir(C^N)$_2$(L)$_2$]$^-$-type Ir complexes. By keeping the cyclometalated ligand C^N as 2-phenylpyridine, Nazeeruddin et al. introduced three pseudohalogens, CN$^-$, NCS$^-$, and NCO$^-$, as spectator ligand L [41]. They investigated the influence of the spectator ligand on the photoluminescent properties of this type of complex {(C$_4$H$_9$)$_4$N[Ir(ppy)$_2$(CN)$_2$], complex **8** in Fig. 10; (C$_4$H$_9$)$_4$N[Ir(ppy)$_2$(SCN)$_2$], complex **40**, and (C$_4$H$_9$)$_4$N [Ir(ppy)$_2$(OCN)$_2$], complex **41**, in Fig. 18)} in 2003 [41]. Introducing a strong ligand field strength spectator ligand such as CN$^-$ increased the gap between LUMO of the phenyl pyridine ligand and metal t_{2g} orbitals, resulting in a blue-shift in the emission spectrum. The gap between the LUMO of the phenyl pyidine ligand and the vacant metal e_g orbitals effectively increased because of the inhibition of nonradiative pathways by the cyanide ligands, which also led to these complexes displaying unusually high phosphorescence quantum yields in solution at room temperature.

While maintaining the spectator ligand L as CN$^-$, Nazeeruddin et al. traced the effect of cyclometalated ligands (**8** in Fig. 10; **42–45** in Fig. 18) in 2008 [85]. They found that introduction of 4-dimethylaminopyridine can destabilize LUMO and HOMO orbitals in parallel, leading to marginal changes in emission spectra, but it increased nonradiative rate constants, which in turn led to a reduction of overall quantum yield. They also found that introducing fluorine atoms on the phenyl

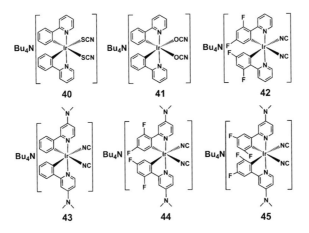

Fig. 18 Chemical structures of complexes **40–45**

Table 3 Photophysical properties of complexes **8** and **40–45** in acetonitrile solution at 298 K

Complex	λ_{max}	Φ_{PL}	τ (µs)
8	470, 502	0.94 ± 0.05	3.14 ± 0.5%
40	506, 520	0.97 ± 0.05	1.43 ± 0.5%
41	538, 560	0.99 ± 0.05	0.85 ± 0.5%
42	460, 485	0.80 ± 0.1	3.28 ± 0.03
43	465, 488, 525sh	0.54 ± 0.1	1.82 ± 0.03
44	451, 471, 525sh	0.62 ± 0.1	1.30 ± 0.03
45	468, 492	0.64 ± 0.1	3.00 ± 0.03

groups led to stabilization of HOMO orbitals. This led to an increase in the HOMO-LUMO gap and was accompanied by a blue-shift in emission spectra, but it did not substantially influence quantum yields because of a slight increase in radiative rate constants or a decrease in overall nonradiative rate constants. Based on these findings, they provided an interesting approach for tuning the phosphorescence wavelength from 470 to 450 nm of anionic Ir complexes, maintaining a high phosphorescence quantum yield by modification with donor and acceptor substituents on the pyridine and phenyl moieties of 2-phenylpyridine (detailed photophysical data are summarized in Table 3).

The photophysical properties of both neutral Ir(N^C^N)(C^N^X) type and ionic $[IrN_{6-n}C_n]^{(3-n)+}$ ($n = 0, 1, 2$) type Ir complexes with tridentate ligands, in particular how the number of cyclometalating carbon atoms in the coordination sphere of the metal ion influences the luminescence, were comprehensively reviewed by Williams in 2008 [31].

4 Applications of Luminescent Iridium Complexes

4.1 Applications in OLEDs

OLEDs are electroluminescent devices in which electrical energy is converted to luminous energy. A typical configuration is depicted in Fig. 19. Two electrodes sandwich one or more layers of the organic functional films. A voltage of 2–10 V is typically applied between the electrodes, and electrons are injected into the LUMO of the organic material from a low work function metallic cathode such as Mg–Ag or Li–Al. Holes are injected into the HOMO of the organic material from a high work function anode such as indium–tin–oxide (ITO). Electrons and holes move towards the middle region of the emitting layer under the influence of the applied field. Energetic electrons can drop into the holes to form singlet and triplet excitons which release their energy as photons escaping through the transparent electrode.

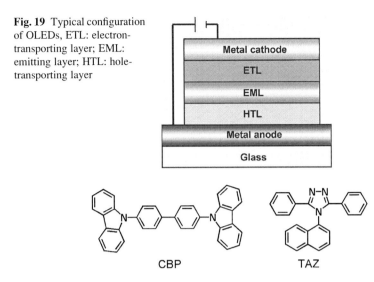

Fig. 19 Typical configuration of OLEDs, ETL: electron-transporting layer; EML: emitting layer; HTL: hole-transporting layer

Fig. 20 Chemical structures of CBP and TAZ

4.1.1 Emitters for Doping OLEDs

Ir complexes have been doped into a charge-transporting host material to optimize device efficiency up to a theoretical limit. The remarkable enhancement in efficiency upon doping has been ascribed to a favorable triplet energy level alignment between host material and Ir complex dopant. Triplet–triplet annihilation, which is a key adverse factor for phosphorescence-based emitters, can also be greatly inhibited by dispersing emitter molecules into the host matrix. Highly efficient red, green, blue (RGB) devices (which are necessary for full-color displays) can be explored by doping different color emission Ir complexes into an appropriate host.

Green-phosphorescent OLED is the best developed RGB device. Cyclometalated *fac*-**1** and its heteroleptic analog such as Ir(ppy)$_2$(acac) (**20** in Fig. 14) have been extensively applied in fabricating green-emitting electroluminescent devices. In 1999, Forrest et al. described high-efficiency OLEDs employing *fac*-**1** as the dopant and CBP (Fig. 20) as the host [2]. The combination of a short triplet lifetime and reasonable photoluminescent efficiency allows *fac*-**1** based OLEDs to achieve peak quantum efficiencies of 8.0%. In 2000 they demonstrated much higher efficiency OLEDs by doping *fac*-**1** into an electron-transport layer host TAZ (Fig. 20) [10]. A maximum external quantum efficiency of 15.4 ± 0.2% was achieved. In 2001, they again demonstrated very high efficiency OLEDs employing **15** doped into TAZ host [20]. A maximum external quantum efficiency of 19.0 ± 1.0% was achieved. The calculated internal quantum efficiency of 87 ± 7% is supported by the observed absence of thermally activated nonradiative loss in the photoluminescent efficiency of **20**. Very high external quantum efficiencies are therefore due to the nearly 100% internal phosphorescence efficiency of the Ir complex coupled with balanced hole and electron injection, as well as triplet exciton confinement within the light-emitting layer.

Compared to green OLEDs, development of red OLEDs has been slower. This is because red emission originated from a smaller energy gap transition, which increased the difficulty of material searching. Many nonradiative pathways originating from strong π–π interaction or charge transfer between ligands may be present, leading to a reduction in efficiency in the OLEDs. The earliest report concerning a red emitting Ir complex as an emitter was by Forrest et al. in 2001 [19]. The device achieved a maximum external quantum efficiency of 7.0 ± 0.5% using a red phosphor Btp$_2$Ir(acac) (**46**, Fig. 21) as the dopant and CBP as the host. In 2003, Tsuboyama et al. fabricated an efficient red OLED device using a novel Ir emitter Ir(piq)$_3$ (**47**, Fig. 21) as a dopant [8]. The maximum external quantum efficiency was 10.3%. Duan et al. synthesized two new orange–red Ir complexes, Ir(DBQ)$_2$(acac) (**21**, Fig. 14) and Ir(MDQ)$_2$(acac) (**48**, Fig. 21) [24]. The devices based on these two complexes emit orange–red light, and the maximum external quantum efficiency is 12%.

Highly efficient blue OLEDs are necessary to realize RGB full-color displays. One of the best known phosphorescence blue emitters is FIrpic (**49**, Fig. 22) [95]. A maximum external electroluminescent quantum efficiency of 7.5 ± 0.8% was obtained by doping **49** into mCP (Fig. 23). The result represents a significant increase in efficiency over a similar device structure based on the host CBP, where the maximum external quantum efficiency was 6.1 ± 0.6%. Data mentioned above indicate that the energy differences in the triplet energies of host and guest materials are very important for confinement of electro-generated triplet excitons on dopant molecules, thus the key factor to achieve highly efficient blue OLEDs is to find appropriate guest and host materials. Based on this guidance, Thompson et al.

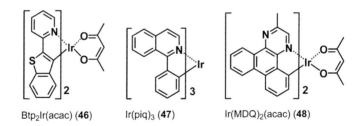

Fig. 21 Chemical structure of red emitters **46–48**

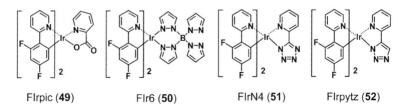

Fig. 22 Chemical structures of blue emitters **49–52**

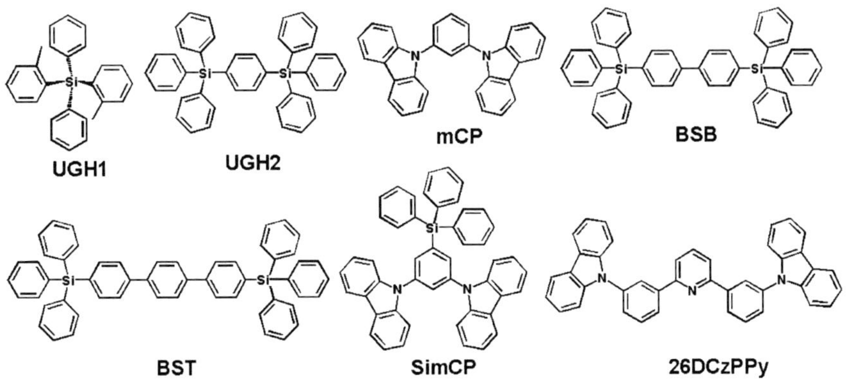

Fig. 23 Chemical structures of some host materials for blue emission iridium complex

demonstrated efficient, deep-blue OLEDs using a charge-trapping phosphorescent guest FIr6 (**50**, Fig. 22) doped in the wide-energy-gap hosts UGH1 and UGH2 (Fig. 23) [25]. Peak quantum efficiencies of 8.8 ± 0.9% in UGH1 and 11.6 ± 1.2% in UGH2 were obtained. In 2005, Yeh et al. reported a series of OLEDs based on FIrN4 (**51**, Fig. 22) or **49** dopant emitters coevaporated with mCP or SimCP (Fig. 23) [48]. The device based on **51** and mCP was found to be one of the bluest OLEDs. The device based on **49** and SimCP achieved maximum external electroluminescent quantum efficiency of 14.4%. Considering UGH1 and UGH2 are less conductive compared with the carbazole derivatives, and that the glass transition temperatures of these two compounds are low for practical application, Lin et al. synthesized two hosts, BSB and BST (Fig. 23), as well as a new blue Ir complex FIrpytz (**52**, Fig. 22) [96]. By using BSB as the host for the blue emitter **52**, highly efficient blue OLEDs with external quantum efficiency of 19.3% was achieved. Su et al. recently reported a unique molecular design strategy of combining a carbazole electron donor with a high triplet energy and a pyridine electron acceptor with high electron affinity to give a novel bipolar host material of 26DCzPPy (Fig. 23) [97]. By using the host for **49**-based blue phosphorescent OLEDs, an external quantum efficiency of 24% was achieved at the practical brightness of 100 cd m^{-2}. Even at a brighter emission of 1,000 cd m^{-2}, an efficiency of 22% was obtained.

Despite the investigation of RGB OLEDs for display applications, Ir complexes have also been used to exploit near-infrared (NIR) OLEDs for optical communication and biomedical application. Jabbour et al. demonstrated the first example of NIR OLEDs fabricated by the cyclometalated Ir complex NIR1 (**53**, Fig. 24) by increasing the size of the cyclometalated ligand π system [98]. The devices exhibited exclusive emission with a peak value at 720 nm, and the external quantum efficiency was nearly 0.1%.

Ir complexes were also used to develop white OLEDs (WOLEDs) for large-scale production of solid-state light sources and backlights in liquid-crystal displays. Several device architectures have been introduced to achieve high brightness and efficiency in WOLEDs. By controlling the recombination current within individual

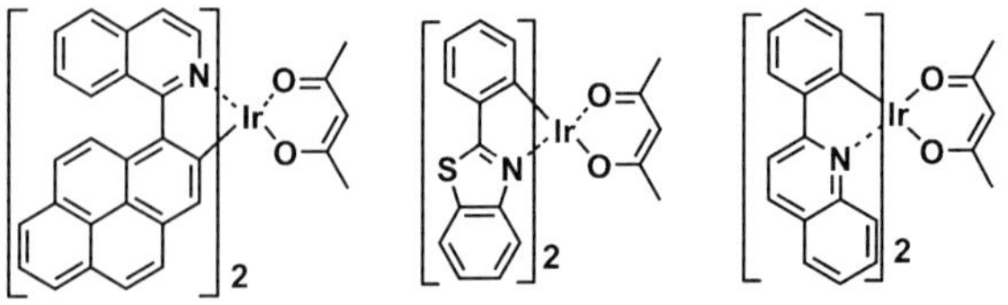

NIR1 (**53**) $Bt_2Ir(acac)$ (**54**) PQIr(acac) (**55**)

Fig. 24 Chemical structures of **53**–**55**

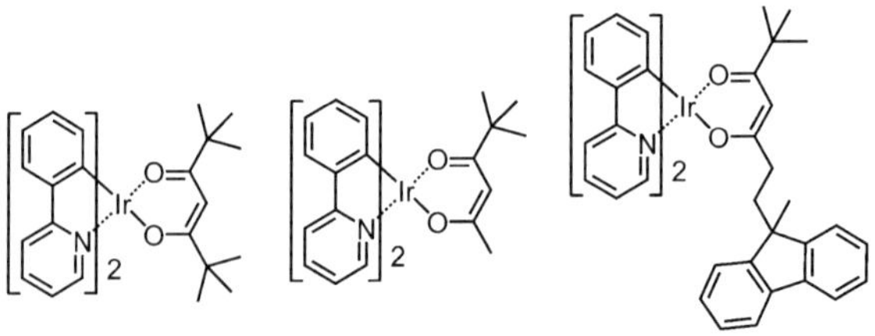

$Ir(ppy)_2(tmd)$ (**56**) $Ir(ppy)_2(dmd)$ (**57**) $Ir(ppy)_2(FBDK)$ (**58**)

Fig. 25 Chemical structures of **56**–**58**

organic layers, emission from red **46**, yellow **54** (Fig. 24), and blue **49** was balanced to obtain white of the desired color purity [99]. The most significant disadvantage of this structure is its relatively high operating voltage due to the combined thicknesses of the many layers used in the emission region. To solve this problem, Forrest et al. mixed red **55** (Fig. 24), green *fac*-**1**, and blue **50** into a wide energy gap UGH2 host to ensure that all emission originates from a single thin layer. The process of direct triplet exciton formation of the **50**-UGH2 system leads to a reduction in operating voltage, and hence an increase in power efficiency up to 42 ± 4 lm W^{-1} was obtained [100].

Doping OLEDs mentioned above are all using fluorescent materials as the host material. Phosphorescent materials can also be used as host material and their performances appear to be significant [50, 101]. Our results proved that highly efficient phosphorescent OLEDs can be pursued using Ir complexes as the host materials [102]. In detail, six devices using **20**, **23** (Fig. 14), **56**, **57** and **58** (Fig. 25) or CBP as the host for an orange–red Ir complex **21** (Fig. 14) were fabricated with an identical configuration. Results show that the devices using Ir complex hosts had better performances than that of the device based on a CBP host. In addition, steric hindrance and exciton-transporting ability of the host were found to be the most important factors for this type of doped system.

4.1.2 Emitters for Non-doped OLEDs

Up to now, most positive results based on Ir complex emitters have been obtained by using a host–guest doped emitter system to improve energy transfer and to avoid triplet–triplet annihilation which occurs when phosphorescent materials are used in OLEDs. Considering that reproducibility of the optimum doping level requires careful manufacturing control, and that long-playing operation of devices may lead to phase separation of guest and host materials, doped OLEDs are relatively more difficult to adapt to practical application than their nondoped counterparts [103]. High-performance phosphorescent OLEDs fabricated by a much simplified, nondoped method are rarely studied, and their typical performances with respect to brightness and efficiency are far from satisfactory [13, 25, 104, 105]. It is therefore desirable to design nondoped electrophosphorescent devices with high brightness and efficiency.

A novel red phosphorescent Ir complex containing carbazole-functionalized β-diketonate **24** (Fig. 14) was designed, synthesized, and characterized in our earlier report [106]. Electrophosphorescent properties of a nondoped device using complex **24** as emitter were examined. The nondoped device achieved maximum lumen efficiency of 3.49 lm W^{-1}. This excellent performance can be attributed mainly to the improved hole-transporting property that benefits exciton transportation. Encouraged by the fact that functionalized β-diketonate can have such an important role in the Ir complex, and with a wish to design high-efficiency nondoped green and blue OLEDs, we synthesized two complexes both containing carbazole-functionalized β-diketonate, **22** and **23**, and subsequently fabricated nondoped devices using these two complexes [107]. Lumen efficiency of the green emission device using **23** as emitter was 4.54 lm W^{-1}, whereas the device based on a blue–green **22** achieved maximum lumen efficiency of 0.51 lm W^{-1}. A very simple device and two double-layer devices **23** were also fabricated, and the designed Ir complex was proved to be with good hole-transporting ability and electron-transporting ability. This study provided a special type of doping technology that should have a more general use in emitter design.

As opposed to neutral Ir complexes, charged Ir complexes are mainly synthesized to develop LECs (another type of OLED) [55, 59, 108–110]. In these devices, electrons and holes injected from two air-stable electrodes into a single layer of organic semiconductor recombine, giving rise to light emission. LECs have advantages over traditional OLEDs: (1) LECs require only a single layer of organic semiconductor, whereas traditional OLEDs require a multilayered structure for charge injection, transport, and light emission, and (2) charge injection in an LEC is insensitive to the work function of the electrode material, thereby permitting use with a wide variety of metals as cathode materials. These suggest that LECs may be a promising alternative for solid-state lighting technologies.

4.2 Biological Labeling Reagents

To investigate the potential of luminescent Ir complexes as biological labeling reagents, Lo et al. [33–36, 61–63] reported a series of $[Ir(C^\wedge N)_2(N^\wedge N-R)](PF_6)$

(R = CHO, NCS and NHCOCH$_2$I) complexes incorporating aldehyde, isothiocyanate, and iodoacetamide groups in the ligands. These Ir complexes can be functionalized as biological labeling reagents due to their ability – covalently or noncovalently – to bind biomolecules. Crosslinked products showed different emission colors than

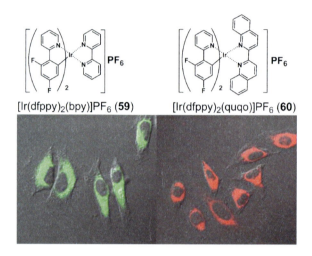

Fig. 26 Chemical structures of **59**, **60**, and their bio-imaging application

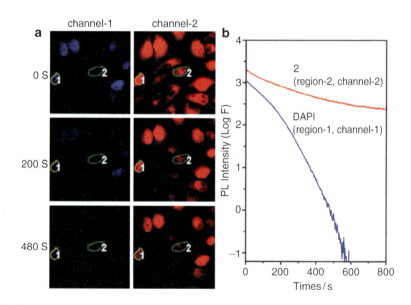

Fig. 27 Comparison of **60** and DAPI for resistance to photobleaching. (**a**) Confocal luminescence images of fixed HeLa cells stained with **60** and DAPI under continuous excitation at 405 nm with different laser scan times (0, 200, 480 s). (**b**) Luminescence decay curves of **60** and DAPI during the same period. The signals of DAPI and **60** were collected from region 1 of channel 1 (460 ± 20 nm) and region 2 of channel 2 (620 ± 20 nm), respectively

their free analogs due to the more hydrophobic environment associated with the protein molecule. A comprehensive review of luminescent Ir complexes used as biological labeling reagents was reported by Lo et al. in 2005 [64].

We recently reported two cationic Ir complexes, **59** (Fig. 26) with bright green emission, and **60** (Fig. 26) with red emission, as phosphorescent dyes for imaging of living cells [111]. These two Ir complexes have advantages when used as bio-imaging agents: exclusive staining in cytoplasm, low cytotoxicity, reduced photobleaching, permeability to cell membranes, and moderate luminescence efficiencies in buffer solution.

As shown in Fig. 27, after continuous excitation at 405 nm for 480 s, luminescence intensity of 4′,6′-diamidino-2-phenylindole dihydrochloride (DAPI) (460 ± 20 nm, region 1, channel 1) decreased to 1% of its initial value (owing to photobleaching). Luminescence intensity of **60** (620 ± 20 nm, region 2, channel 2) stayed at essentially one-eighth of the original value during the same period of excitation. This result establishes that the Ir complex show reduced photobleaching and higher photostability than the organic dye. These findings open-up interesting possibilities for using luminescent Ir complexes for imaging of living cells. Investigation of other Ir complexes used in this field will be published in a forthcoming article.

4.3 Sensitizer of Lanthanide Luminescence

Lanthanide ions such as europium (Eu^{III}), neodymium (Nd^{III}), ytterbium (Yb^{III}), and erbium (Er^{III}) are becoming increasingly important in applications such as OLEDs [112–115], optical communication [116, 117], medical imaging, and biological labeling [118, 119] due to their excellent luminescence properties originating from f–f transitions. These ions show little or no absorption in the visible region of the spectrum, and often require application of strongly absorbing "antennae" for light harvesting to obtain efficient photoluminescence [120]. Many organic chromophores having high absorbing ability were therefore introduced into lanthanide complexes. The use of strongly-absorbing d-block chromophores (e.g., Ir complexes) as sensitizers has attracted increasing attention [65–67].

The first example of sensitization of Eu^{III} emission through an Ir complex was reported by De Cola et al. in 2005 (Fig. 28) [67]. Energy transfer from the Ir fragments to the Eu^{III} center is incomplete, leading to emission of a composite white light instead of Eu^{III}-based emission of pure red light.

To make energy transfer more efficient and obtain pure-red emission from Eu^{III}, a novel ligand with four coordination sites was designed as a bridge to link the Ir^{III} center and the Eu^{III} center; and a new Ir^{III} complex Ir(dfppy)$_2$(phen5f) [dfppy represents 2-(4′,6′-difluorophenyl)-pyridinato-N,$C^{2′}$, phen5f denotes 4,4,5,5,5-pentafluoro-1-(1′,10′-phenanthrolin-2′-yl)-pentane-1,3-dionate] was obtained [65]. By using the "complexes as ligands" approach, the novel bimetallic complex

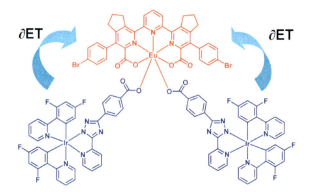

Fig. 28 The first example of sensitization of Eu[III] emission through an iridium complex

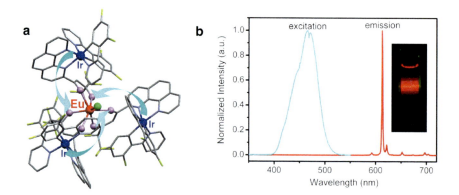

Fig. 29 (a) The crystal structure of "Ir₃Eu" assembly. (b) Pure-red emission of the "Ir₃Eu" assembly in EtOH (1×10^{-3} M) upon excitation of a 532-nm laser

{[(dfppy)₂Ir(μ-phen5f)]₃EuCl}Cl₂ was synthesized, and efficient pure-red luminescence from Eu[III] was sensitized by ³MLCT energy from the Ir[III] moiety (Fig. 29). The excitation window can extend up to 530 nm due to the introduction of the d-block metal moiety, so this bimetallic complex can readily emit red light under sunlight irradiation.

As important phosphorescent materials, many Ir[III] complexes with different energies of the lowest excited states have been extensively investigated by modifying cyclometalated ligands. This is a good strategy to seek more suitable Ir[III] complexes to sensitize NIR lanthanide ions. Our research group has changed the cyclometalated ligand from dfppy to ppy to reduce the triplet energy level of the whole d-block complex-ligand and make it more suitable for NIR Ln[III]. Another Ir[III] complex Ir(ppy)₂(phen5f) was introduced to form Ir₂Ln (Ln = Nd, Yb, Er) arrays. NIR emission upon photoexcitation of the Ir[III]-centered antenna

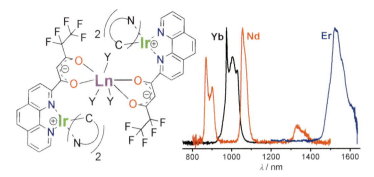

Fig. 30 Structures of Ir$_2$Ln (Ln = Nd, Yb, Er) arrays and corresponding near-infrared emissions. C^N denotes cyclometalated ligand ppy, and Y denotes bidentate nitrate anion

chromophore was successfully obtained (Fig. 30) [121]. An article involving sensitized NIR emission from Yb via direct energy transfer from Ir in a heterometallic neutral complex was published at virtually the same time by De Cola et al. [66].

4.4 Sensor Applications

Photoluminescent materials in which luminescence output can be modified by interaction with a substrate are being extensively investigated for use as sensors. Luminescent Ir complexes have been used as oxygen sensors, homocysteine sensors, metal cation sensors, and volatile organic compound sensors because of their rich photophysical properties.

4.4.1 Oxygen Sensor

Luminescence-based oxygen sensors work on the principle of luminescence quenching by oxygen; the excited luminophore enables efficient energy transfer to the triplet ground state of molecular oxygen, resulting in a nonradiative luminophore and formation of singlet oxygen [122]. Ir complexes are attractive candidates as novel luminescence-based oxygen-sensing materials (primarily because Ir complexes are the best known luminescent metal complexes). They can be excited with visible light, and their excited state is mainly a mixed ^{3}MLCT and ^{3}LC level, which is prone to quenching by molecular oxygen.

The first example of an Ir complex-based oxygen sensor was reported by Donckt et al. in 1994 [123]. They embedded *fac*-**1** in polystyrene and studied the luminescence properties of the system for use as an oxygen sensor to avoid self-quenching of *fac*-**1** at higher concentration. In 1996, DiMarco et al. [124] reported another Ir complex [Ir(ppy)$_2$(dpt-NH$_2$)](PF$_6$) (where dpt-NH$_2$ = 4-amino-3,5-di-2-pyridyl-4*H*-1,2,4-triazole), immobilized in a polymerized poly-(ethylene-glycol) ethyl ether methacrylate (pPEGMA) matrix. The system exhibited excellent properties as a quenchometric oxygen sensor (e.g., linear Stern–Volmer plot, photostability, thermal stability, and reproducibility). They extended their study in 1998 and reported the oxygen sensor ability of a series of mononuclear and dinuclear cyclometalated Ir complexes immobilized in pPEGMA matrixes [125]. The sensitivity and response of the sensor system was affected not only by the lifetime of the Ir complex and the permeability of matrices, but also by the size and charge of the Ir complex. This is important for the design of new solid-state luminescent sensors with improved performance. Efforts were subsequently focused on covalently binding the luminophore to different polymeric hosts, as well as adjusting the structure of the Ir complex [11, 68, 70, 122, 126].

4.4.2 Homocysteine Sensor

Homocysteine has a unique role within physiologic matrices because it is an important amino acid containing a free thiol moiety. Detection of homocysteine from other amino acids is therefore important. A selective phosphorescence chemosensor for this aim was developed based on the reaction shown in Fig. 31 [127]. Upon addition of homocysteine to a semiaqueous solution of **61**, a color change from orange to yellow and a luminescent variation from deep red to green were evident to the naked eye. This can be attributed to formation of a thiazinane group by selective reaction of the aldehyde group of **61** with homocysteine.

4.4.3 Metal Cation Sensor

Binding an appropriate metal ion to the ligand may lead to sufficient changes in luminescence because the photophysical properties of the Ir complex are influenced by the structure or surrounding environment of the organic ligands. In 2006, Ho et al. [128] presented a novel system in which an azacrown receptor was attached to the pyridyl pyrazolate chelate of a heteroleptic Ir complex. Photophysical study showed that phosphorescence was gradually blue-shifted from 560 to 520 nm, and was accompanied by an increase of emission intensity upon addition of Ca^{2+}, which made the complex a highly sensitive phosphorescence probe. In 2007, Schmittel

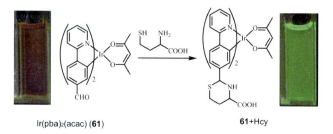

Fig. 31 Luminescent iridium complex-based sensor for homocysteine

et al. [129] synthesized **62** by introducing a crown group. The complex exhibited selective binding properties toward Ag^+ and Hg^{2+} in aqueous media, accompanied by characteristic luminescence responses. As shown in Fig. 32, luminescence was enhanced by >10 times in the presence of excess Ag^+; this is because the diimine ligand becomes an increasingly better acceptor by binding Ag^+. The emission quenched about 80% upon addition of Hg^{2+}, which may be explained by competition between emission-enhancing effects (decrease of the electron-donating ability of nitrogen upon Hg^{2+} binding) and the electron- or energy transfer-quenching effects of unbound Hg^{2+} ions. Ho et al. [130] demonstrated the concept of Pb^{2+} cation-sensing using the emissive Ir complex, which is based on the associated reduction of phosphorescence at room temperature upon chelate interaction between the Ir complex and metal analyte.

4.4.4 Acetonitrile or Propiononitrile Vapor Sensor

Sensor applications involving significant color and/or luminescence efficiency tuning mentioned above are usually based on structural or coordinated environment changes in the cyclometalated or ancillary ligand. We recently found the first example of an Ir complex PIrqnx (**63**) having a unique, fast vapochromic and vapoluminescent behavior towards acetonitrile or propiononitrile vapor based on molecular packing transformation (Fig. 33) [131]. Complex **63** exists as black

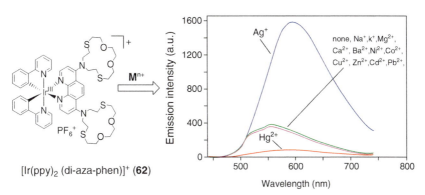

Fig. 32 Iridium-based chemosensors for Ag^+ and Hg^{2+} in aqueous media

Fig. 33 An iridium complex that can detect acetonitrile or propiononitrile vapor

and red forms (visible to the naked eye) in the solid state. The black form can be transformed to the red form upon exposure to acetonitrile or propiononitrile vapor, but no response was observed when it was exposed to other volatile organic compounds. Crystallographic and DFT studies indicated that the black form adsorbed acetonitrile vapor first because of a porous packing structure, then the red form was formed with different color and luminescence properties induced by weak intermolecular interactions (e.g., hydrogen bonding, $\pi-\pi$ interactions).

5 Summaries and Outlook

As can be seen from the evolution of Ir complexes discussed in this chapter, chemical syntheses are carried out to coordinate different ligand structures to the metal center to control the excited state. Ir complexes are therefore tailored to express specific luminescent properties, and are being explored for many applications (e.g., OLEDs, biological labeling reagents, chemosensors). And with the efforts of research groups investigating luminescent Ir complexes, we believe the future is very bright indeed.

Acknowledgments The authors acknowledge financial support from the National Basic Research Program of China (2006CB601103) and the National Natural Science Foundation of China (20021101, 20423005, 20471004, 50372002, and 20671006).

References

1. Baldo MA, O'Brien DF, You Y, Shoustikov A, Sibley S, Thompson ME, Forrest SR (1998) Nature 395:151–154
2. Baldo MA, Lamansky S, Burrows PE, Thompson ME, Forrest SR (1999) Appl Phys Lett 75:4–6
3. Watts RJ, Houten JV (1974) J Am Chem Soc 96:4334–4335
4. Dedeian K, Djurovich PI, Garces FO, Carlson G, Watts RJ (1991) Inorg Chem 30:1685–1687
5. Lowry MS, Goldsmiths JI, Slinker JD, Rohl R, Pascal RA, Malliaras GG, Bernhard S (2005) Chem Mater 17:5712–5719
6. Lamansky S, Djurovich P, Murphy D, Abdel-Razzaq F, Lee HE, Adachi C, Burrows PE, Forrest SR, Thompson ME (2001) J Am Chem Soc 123:4304–4312
7. Zhou GJ, Wong WY, Yao B, Xie ZY, Wang LX (2007) Angew Chem Int Ed 46:1149–1151
8. Tsuboyama A, Iwawaki H, Furugori M, Mukaide T, Kamatani J, Igawa S, Moriyama T, Miura S, Takiguchi T, Okada S, Hoshino M, Ueno K (2003) J Am Chem Soc 125:12971–12979
9. Okada S, Okinaka K, Iwawaki H, Furugori M, Hashimoto M, Mukaide T, Kamatani J, Igawa S, Tsuboyama A, Takiguchi T, Ueno K (2005) Dalton Trans:1583–1590
10. Adachi C, Baldo MA, Forrest SR, Thompson ME (2000) Appl Phys Lett 77:904–906
11. Amao Y, Ishikawa Y, Okura I (2001) Anal Chim Acta 445:177–182

12. Blumstengel S, Meinardi F, Tubino R, Gurioli M, Jandke M, Strohriegl P (2001) J Chem Phys 115:3249–3255
13. Grushin VV, Herron N, LeCloux DD, Marshall WJ, Petrov VA, Wang Y (2001) Chem Commun 1494–1495
14. Adamovich V, Brooks J, Tamayo A, Alexander AM, Djurovich PI, D'Andrade BW, Adachi C, Forrest SR, Thompson ME (2002) New J Chem 26:1171–1178
15. Tamayo AB, Alleyne BD, Djurovich PI, Lamansky S, Tsyba I, Ho NN, Bau R, Thompson ME (2003) J Am Chem Soc 125:7377–7387
16. King KA, Spellane PJ, Watts RJ (1985) J Am Chem Soc 107:1431–1432
17. Colombo MG, Brunold TC, Riedener T, Gudel HU, Fortsch M, Burgi HB (1994) Inorg Chem 33:545–550
18. Brunner K, van Dijken A, Borner H, Bastiaansen J, Kiggen NMM, Langeveld BMW (2004) J Am Chem Soc 126:6035–6042
19. Adachi C, Baldo MA, Forrest SR, Lamansky S, Thompson ME, Kwong RC (2001) Appl Phys Lett 78:1622–1624
20. Adachi C, Baldo MA, Thompson ME, Forrest SR (2001) J Appl Phys 90:5048–5051
21. Adachi C, Kwong RC, Djurovich P, Adamovich V, Baldo MA, Thompson ME, Forrest SR (2001) Appl Phys Lett 79:2082–2084
22. Ikai M, Tokito S, Sakamoto Y, Suzuki T, Taga Y (2001) Appl Phys Lett 79:156–158
23. Lamansky S, Djurovich P, Murphy D, Abdel-Razzaq F, Kwong R, Tsyba I, Bortz M, Mui B, Bau R, Thompson ME (2001) Inorg Chem 40:1704–1711
24. Duan JP, Sun PP, Cheng CH (2003) Adv Mater 15:224–228
25. Holmes RJ, D'Andrade BW, Forrest SR, Ren X, Li J, Thompson ME (2003) Appl Phys Lett 83:3818–3820
26. Su YJ, Huang HL, Li CL, Chien CH, Tao YT, Chou PT, Datta S, Liu RS (2003) Adv Mater 15:884–888
27. Tokito S, Iijima T, Tsuzuki T, Sato F (2003) Appl Phys Lett 83:2459–2461
28. Coppo P, Plummer EA, De Cola L (2004) Chem Commun 1774–1775
29. Wilkinson AJ, Puschmann H, Howard JAK, Foster CE, Williams JAG (2006) Inorg Chem 45:8685–8699
30. Wilkinson AJ, Goeta AE, Foster CE, Williams JAG (2004) Inorg Chem 43:6513–6515
31. Williams JAG, Wilkinson AJ, Whittle VL (2008) Dalton Trans 2081–2099
32. Neve F, Crispini A (2000) Eur J Inorg Chem 1039–1043
33. Lo KKW, Ng DCM, Chung CK (2001) Organometallics 20:4999–5001
34. Lo KKW, Chung CK, Ng DCM, Zhu NY (2002) New J Chem 26:81–88
35. Lo KKW, Chung CK, Lee TKM, Lui LH, Tsang KHK, Zhu NY (2003) Inorg Chem 42:6886–6897
36. Lo KKW, Chung CK, Zhu NY (2003) Chem Eur J 9:475–483
37. Colombo MG, Gudel HU (1993) Inorg Chem 32:3081–3087
38. Colombo MG, Hauser A, Gudel HU (1993) Inorg Chem 32:3088–3092
39. Ohsawa Y, Sprouse S, King KA, Dearmond MK, Hanck KW, Watts RJ (1987) J Phys Chem 91:1047–1054
40. Maestri M, Sandrini D, Balzani V, Maeder U, Vonzelewsky A (1987) Inorg Chem 26:1323–1327
41. Nazeeruddin MK, Humphry-Baker R, Berner D, Rivier S, Zuppiroli L, Graetzel M (2003) J Am Chem Soc 125:8790–8797
42. Whittle VL, Williams JAG (2008) Inorg Chem 47:6596–6607
43. Goodall W, Wild K, Arm KJ, Williams JAG (2002) J Chem Soc Perkin Trans 2:1669–1681
44. Leslie W, Batsanov AS, Howard JAK, Williams JAG (2004) Dalton Trans 623–631
45. Inomata H, Goushi K, Masuko T, Konno T, Imai T, Sasabe H, Brown JJ, Adachi C (2004) Chem Mater 16:1285–1291
46. Paulose B, Rayabarapu DK, Duan JP, Cheng CH (2004) Adv Mater 16:2003–2007

47. Rayabarapu DK, Paulose B, Duan JP, Cheng CH (2005) Adv Mater 17:349–353
48. Yeh SJ, Wu MF, Chen CT, Song YH, Chi Y, Ho MH, Hsu SF, Chen CH (2005) Adv Mater 17:285–289
49. Guan M, Chen ZQ, Bian ZQ, Liu ZW, Gong ZL, Baik W, Lee HJ, Huang CH (2006) Org Electron 7:330–336
50. Tsuzuki T, Tokito S (2007) Adv Mater 19:276–280
51. Ho CL, Wong WY, Gao ZQ, Chen CH, Cheah KW, Yao B, Xie ZY, Wang Q, Ma DG, Wang LA, Yu XM, Kwok HS, Lin ZY (2008) Adv Funct Mater 18:319–331
52. Bolink HJ, Cappelli L, Coronado E, Gratzel M, Orti E, Costa RD, Viruela PM, Nazeeruddin MK (2006) J Am Chem Soc 128:14786–14787
53. Bolink HJ, Cappelli L, Coronado E, Parham A, Stossel P (2006) Chem Mater 18:2778–2780
54. Chen FC, Yang Y, Pei Q (2002) Appl Phys Lett 81:4278–4280
55. Su HC, Chen HF, Fang FC, Liu CC, Wu CC, Wong KT, Liu YH, Peng SM (2008) J Am Chem Soc 130:3413–3419
56. Zeng XS, Tavasli M, Perepichka IE, Batsanov AS, Bryce MR, Chiang CJ, Rothe C, Monkman AP (2008) Chem Eur J 14:933–943
57. Bolink HJ, Cappelli L, Cheylan S, Coronado E, Costa RD, Lardies N, Nazeeruddin MK, Orti E (2007) J Mater Chem 17:5032–5041
58. Dragonetti C, Falciola L, Mussini P, Righetto S, Roberto D, Ugo R, Valore A, De Angelis F, Fantacci S, Sgamellotti A, Ramon M, Muccini M (2007) Inorg Chem 46:8533–8547
59. Tamayo AB, Garon S, Sajoto T, Djurovich PI, Tsyba IM, Bau R, Thompson ME (2005) Inorg Chem 44:8723–8732
60. Nazeeruddin MK, Wegh RT, Zhou Z, Klein C, Wang Q, De Angelis F, Fantacci S, Gratzel M (2006) Inorg Chem 45:9245–9250
61. Lo KKW, Zhang KY, Leung SK, Tang MC (2008) Angew Chem Int Ed 47:2213–2216
62. Lo KKW, Zhang KY, Chung CK, Kwok KY (2007) Chem Eur J 13:7110–7120
63. Lo KKW, Chung CK, Zhu NY (2006) Chem Eur J 12:1500–1512
64. Lo KKW, Hui WK, Chung CK, Tsang KHK, Ng DCM, Zhu NY, Cheung KK (2005) Coord Chem Rev 249:1434–1450
65. Chen FF, Bian ZQ, Liu ZW, Nie DB, Chen ZQ, Huang CH (2008) Inorg Chem 47:2507–2513
66. Mehlstaubl M, Kottas GS, Colella S, De Cola L (2008) Dalton Trans 2385–2388
67. Coppo P, Duati M, Kozhevnikov VN, Hofstraat JW, De Cola L (2005) Angew Chem Int Ed 44:1806–1810
68. DeRosa MC, Hodgson DJ, Enright GD, Dawson B, Evans CEB, Crutchley RJ (2004) J Am Chem Soc 126:7619–7626
69. Borisov SM, Klimant I (2007) Anal Chem 79:7501–7509
70. Fernandez-Sanchez JF, Roth T, Cannas R, Nazeeruddin MK, Spichiger S, Graetzel M, Spichiger-Keller UE (2007) Talanta 71:242–250
71. Medina-Castillo AL, Fernandez-Sanchez JF, Klein C, Nazeeruddin MK, Segura-Carretero A, Fernandez-Gutierrez A, Graetzel M, Spichiger-Keller UE (2007) Analyst 132:929–936
72. Lo SC, Male NAH, Markham JPJ, Magennis SW, Burn PL, Salata OV, Samuel IDW (2002) Adv Mater 14:975–979
73. Markham JPJ, Samuel IDW, Lo SC, Burn PL, Weiter M, Bassler H (2004) J Appl Phys 95:438–445
74. Lo SC, Richards GJ, Markham JPJ, Namdas EB, Sharma S, Burn PL, Samuel IDW (2005) Adv Funct Mater 15:1451–1458
75. Zhou GJ, Wong WY, Yao B, Xie Z, Wang L (2008) J Mater Chem 18:1799–1809
76. Li XH, Chen Z, Zhao Q, Shen L, Li FY, Yi T, Cao Y, Huang CH (2007) Inorg Chem 46:5518–5527
77. Hwang SH, Shreiner CD, Moorefield CN, Newkome GR (2007) New J Chem 31:1192–1217
78. Lo SC, Burn PL (2007) Chem Rev 107:1097–1116
79. Nonoyama M (1974) Bull Chem Soc Jpn 47:767–768

80. Li J, Djurovich PI, Alleyne BD, Tsyba I, Ho NN, Bau R, Thompson ME (2004) Polyhedron 23:419–428
81. Song YH, Yeh SJ, Chen CT, Chi Y, Liu CS, Yu JK, Hu YH, Chou PT, Peng SM, Lee GH (2004) Adv Funct Mater 14:1221
82. Chen XW, Liao JL, Liang YM, Ahmed MO, Tseng HE, Chen SA (2003) J Am Chem Soc 125:636–637
83. Nonoyama M (1974) J Organomet Chem 82:271–276
84. Zhao Q, Liu SJ, Shi M, Wang CM, Yu MX, Li L, Li FY, Yi T, Huang CH (2006) Inorg Chem 45:6152–6160
85. Di Censo D, Fantacci S, De Angelis F, Klein C, Evans N, Kalyanasundaram K, Bolink HJ, Gratzel M, Nazeeruddin MK (2008) Inorg Chem 47:980–989
86. Ayala NP, Flynn CM, Sacksteder L, Demas JN, Degraff BA (1990) J Am Chem Soc 112:3837–3844
87. Baranoff E, Collin JP, Flamigni L, Sauvage JP (2004) Chem Soc Rev 33:147–155
88. Douglas B, McDaniel D, Alexander J (1994) Concepts and models in inorganic chemistry, 3rd edn. Wiley, New York
89. Liu ZW, Nie DB, Bian ZQ, Chen FF, Lou B, Bian J, Huang CH (2008) ChemPhysChem 9:634–640
90. You Y, Seo J, Kim SH, Kim KS, Ahn TK, Kim D, Park SY (2008) Inorg Chem 47:1476–1487
91. You Y, Kim KS, Ahn TK, Kim D, Park SY (2007) J Phys Chem C 111:4052–4060
92. You YM, Park SY (2005) J Am Chem Soc 127:12438–12439
93. Hay PJ (2002) J Phys Chem A 106:1634–1641
94. De Angelis F, Fantacci S, Evans N, Klein C, Zakeeruddin SM, Moser JE, Kalyanasundaram K, Bolink HJ, Gratzel M, Nazeeruddin MK (2007) Inorg Chem 46:5989–6001
95. Holmes RJ, Forrest SR, Tung YJ, Kwong RC, Brown JJ, Garon S, Thompson ME (2003) Appl Phys Lett 82:2422–2424
96. Lin JJ, Liao WS, Huang HJ, Wu FI, Cheng CH (2008) Adv Funct Mater 18:485–491
97. Su SJ, Sasabe H, Takeda T, Kido J (2008) Chem Mater 20:1691–1693
98. Williams EL, Li J, Jabbour GE (2006) Appl Phys Lett 89:083506
99. D'Andrade BW, Thompson ME, Forrest SR (2002) Adv Mater 14:147–151
100. D'Andrade BW, Holmes RJ, Forrest SR (2004) Adv Mater 16:624–628
101. Kwong RC, Lamansky S, Thompson ME (2000) Adv Mater 12:1134–1138
102. Liu ZW, Bian ZQ, Hao F, Nie DB, Ding F, Chen ZQ, Huang CH (2009) Org Electron. doi:10.1016/j.orgel.2008.11.013
103. Gong JR, Wan LJ, Lei SB, Bai CL, Zhang XH, Lee ST (2005) J Phys Chem B 109:1675–1682
104. Song YH, Yeh SJ, Chen CT, Chi Y, Liu CS, Yu JK, Hu YH, Chou PT, Peng SM, Lee GH (2004) Adv Funct Mater 14:1221–1226
105. Wang Y, Herron N, Grushin VV, LeCloux D, Petrov V (2001) Appl Phys Lett 79:449–451
106. Liu ZW, Guan M, Bian ZQ, Nie DB, Gong ZL, Li ZB, Huang CH (2006) Adv Funct Mater 16:1441–1448
107. Liu ZW, Bian ZQ, Ming L, Ding F, Shen HY, Nie DB, Huang CH (2008) Org Electron 9:171–182
108. Slinker JD, Koh CY, Malliaras GG, Lowry MS, Bernhard S (2005) Appl Phys Lett 86:173506
109. Slinker JD, Gorodetsky AA, Lowry MS, Wang JJ, Parker S, Rohl R, Bernhard S, Malliaras GG (2004) J Am Chem Soc 126:2763–2767
110. Lowry MS, Goldsmith JI, Slinker JD, Rohl R, Pascal RA, Malliaras GG, Bernhard S (2005) Chem Mater 17:5712–5719
111. Yu MX, Zhao Q, Shi LX, Li FY, Zhou ZG, Yang H, Yia T, Huang CH (2008) Chem Commun 2115–2117

112. Adachi C, Baldo MA, Forrest SR (2000) J Appl Phys 87:8049–8055
113. Hong ZR, Liang CJ, Li RG, Li WL, Zhao D, Fan D, Wang DY, Chu B, Zang FX, Hong LS, Lee ST (2001) Adv Mater 13:1241–1245
114. Sun M, Xin H, Wang KZ, Zhang YA, Jin LP, Huang CH (2003) Chem Commun 702–703
115. Liang FS, Zhou QG, Cheng YX, Wang LX, Ma DG, Jing XB, Wang FS (2003) Chem Mater 15:1935–1937
116. Kang TS, Harrison BS, Bouguettaya M, Foley TJ, Boncella JM, Schanze KS, Reynolds JR (2003) Adv Funct Mater 13:205–210
117. Kawamura Y, Wada Y, Hasegawa Y, Iwamuro M, Kitamura T, Yanagida S (1999) Appl Phys Lett 74:3245–3247
118. Motson GR, Fleming JS, Brooker S (2004) Potential applications for the use of lanthanide complexes as luminescent biolabels. In: Advances in inorganic chemistry: including bio-inorganic studies, vol 55. Academic, London, pp 361–432
119. Bunzli JCG, Piguet C (2005) Chem Soc Rev 34:1048–1077
120. Werts MHV, Jukes RTF, Verhoeven JW (2002) Phys Chem Chem Phys 4:1542–1548
121. Chen FF, Bian ZQ, Lou B, Ma E, Liu ZW, Nie DB, Chen ZQ, Bian J, Chen ZN, Huang CH (2008) Dalton Trans 41:5577–5583
122. DeRosa MC, Mosher PJ, Evans CEB, Crutchley RJ (2003) Macromol Symp 196:235–248
123. Vanderdonckt E, Camerman B, Hendrick F, Herne R, Vandeloise R (1994) Bull Soc Chim Belg 103:207–211
124. DiMarco G, Lanza M, Pieruccini M, Campagna S (1996) Adv Mater 8:576–580
125. Di Marco G, Lanza M, Mamo A, Stefio I, Di Pietro C, Romeo G, Campagna S (1998) Anal Chem 70:5019–5023
126. DeRosa MC, Mosher PJ, Yap GPA, Focsaneanu KS, Crutchley RJ, Evans CEB (2003) Inorg Chem 42:4864–4872
127. Chen HL, Zhao Q, Wu YB, Li FY, Yang H, Yi T, Huang CH (2007) Inorg Chem 46:11075–11081
128. Ho ML, Hwang FM, Chen PN, Hu YH, Cheng YM, Chen KS, Lee GH, Chi Y, Chou PT (2006) Org Biomol Chem 4:98–103
129. Schmittel M, Lin HW (2007) Inorg Chem 46:9139–9145
130. Ho ML, Cheng YM, Wu LC, Chou PT, Lee GH, Hsu FC, Chi Y (2007) Polyhedron 26:4886–4892
131. Liu ZW, Bian ZQ, Bian J, Li ZD, Nie DB, Huang CH (2008) Inorg Chem 47:8025–8030

Chromo- and Fluorogenic Organometallic Sensors

Nicholas C. Fletcher and M. Cristina Lagunas

Abstract Compounds that change their absorption and/or emission properties in the presence of a target ion or molecule have been studied for many years as the basis for optical sensing. Within this group of compounds, a variety of organometallic complexes have been proposed for the detection of a wide range of analytes such as cations (including H^+), anions, gases (e.g. O_2, SO_2, organic vapours), small organic molecules, and large biomolecules (e.g. proteins, DNA). This chapter focuses on work reported within the last few years in the area of organometallic sensors. Some of the most extensively studied systems incorporate metal moieties with intense long-lived metal-to-ligand charge transfer (MLCT) excited states as the *reporter* or *indicator* unit, such as *fac*-tricarbonyl Re(I) complexes, cyclometallated Ir(III) species, and diimine Ru(II) or Os(II) derivatives. Other commonly used organometallic sensors are based on Pt-alkynyls and ferrocene fragments. To these *reporters*, an appropriate *recognition* or analyte-binding unit is usually attached so that a detectable modification on the colour and/or the emission of the complex occurs upon binding of the analyte. Examples of recognition sites include macrocycles for the binding of cations, H-bonding units selective to specific anions, and DNA intercalating fragments. A different approach is used for the detection of some gases or vapours, where the sensor's response is associated with changes in the crystal packing of the complex on absorption of the gas, or to direct coordination of the analyte to the metal centre.

Keywords Molecular Sensors, Molecular recognition, Organometallic, Fluorescence, UV-vis spectroscopy, Colorimetry

N.C. Fletcher and M.C. Lagunas (✉)
Queen's University Belfast, School of Chemistry and Chemical Engineering, David Keir Building, Stranmillis Road, Belfast BT9 5AG, Northern Ireland, UK
e-mail: c.lagunas@qub.ac.uk, n.fletcher@qub.ac.uk

Contents

1 Introduction 144
2 Cation Sensors 145
   2.1 Rhenium(I) Carbonyl Complexes 145
   2.2 Ruthenium(II) and Osmium(II) Complexes 147
   2.3 Iridium(III) Cyclometallated Complexes 147
   2.4 Ferrocene-Based Sensors 148
   2.5 Platinum(II) Complexes 150
   2.6 Gold(I) Complexes 151
3 Anion Sensors 152
   3.1 Rhenium(I) Carbonyl Complexes 152
   3.2 Platinum(II), Ruthenium(II) and Iridium(III) Complexes 154
   3.3 Ferrocene-Based Sensors 155
4 Small Molecule Detection 156
   4.1 Detection of Gases and Volatile Organic Compounds 156
   4.2 Detection of Small Organic Molecules in Solution 159
5 Sensing of Large Biomolecules 161
   5.1 Detection of DNA 161
   5.2 Detection of Proteins 163
6 Conclusions and Outlook 166
References 166

1 Introduction

The colorimetric and fluorescent detection of localized molecular environments has become pivotal in the development of working sensor devices [1]. The diimine complexes of the late transition metals, many of which possess long lived excited states, have been in the vanguard of sensor design. Simple variation in their substituents not only allows the addition of a large variety of recognition sites but permits the photophysical properties to be tuned [2, 3]. As a result, a number of organometallic species of these relatively inert metals have found potential application in the detection of a wide variety of analytes, ranging from cations (including H^+), anions, gases and small organic molecules, all the way through to large biomolecules such as DNA.

As a general rule, a molecular sensor is comprised of two sections, usually located in two distinct domains. One part is employed in the *recognition* of the target analyte. Through the considerable efforts within the fields of supramolecular and host–guest chemistry, suitable recognition units have been designed for the detection of a wide variety of materials. The second key unit is the *indicator* or *reporter* that, upon a successful recognition event, undergoes a detectable change such as in the colour, the fluorescence or even the redox potential. One of the most extensively studied systems is the tris(2,2′-bipyridyl)ruthenium(II) cation as a consequence of its unique combination of chemical stability and emissive behaviour. There are numerous examples where this simple moiety has been incorporated into larger sensing devices, and will not be considered further in this chapter [4]. However, it is noted that many of the ligand systems explored with this stalwart of coordination chemistry have also been incorporated into a range of other organometallic luminescent platforms, such as the *fac*-tricarbonyl rhenium(I) moiety, which possesses a long lived triplet metal-to-ligand charge transfer (^{3}MLCT) state. Cyclometallated iridium(III) complexes have also drawn considerable recent attention [5], as they offer the possibilities of tuning emissive states by ligand modification opening routes to red, green and even blue fluorescent materials, as well as larger quantum yields in comparison to the analogous Ru(II) complexes. Attention is also drawn to the acetylene complexes of platinum which have undergone considerable developments in recent years [6].

The objective of this chapter is to draw attention to a number of recently reported organometallic complexes which detectably change their optical properties in the presence of a target analyte. While every effort has been made to ensure that the chapter is as inclusive as possible, a degree of selectivity has had to be undertaken.

2 Cation Sensors

2.1 *Rhenium(I) Carbonyl Complexes*

Tricarbonyl Re(I) complexes $[Re(CO)_3L]^+$, where L is usually a bipyridine ligand, are often used as optical sensors based on their low-lying MLCT excited states $[d\pi(Re) \rightarrow \pi^*(L)]$. A common strategy for cation sensing is the attachment to L of pendant macrocycles, such as crown ethers. In order to ensure a good electronic communication between the cation-binding macrocycle and the reporter Re centre, Lazarides et al. [7] have recently developed ditopic ligands derived from 5,6-dihydroxy-1,10-phenanthroline where pendant crown ether groups are directly linked to the phenanthroline fragment (**1a**, **1b**; Fig. 1). A limitation of the system was found to be the poor donor ability of the O atoms directly connected to the electron-withdrawing phenanthroline–Re moiety, which resulted in unexpected low binding constants for K^+. However, **1a** and **1b** both showed strong binding to Ba^{2+}, which was easily detectable through the quenching of the ^{3}MLCT luminescence band (and a slight red-shift). The incorporation of soft donor atoms such as S or Se in the pendant groups allows for specific binding of soft metal cations. For example, Yam et al. [8] have achieved highly selective sensing of Pb^{2+} and Hg^2 using complexes **1c** and **1d**, respectively (Fig. 1). In this case, spectral changes were observed both in the UV–Vis and luminescence spectra of the compounds. In particular, a significant enhancement of the ^{3}MLCT luminescence intensity (and a slight red-shift) was observed upon addition of the metal cations in acetonitrile. An alternative approach to facilitate effective electron communication between the pendant group and the reporter unit is the incorporation of conjugated spacers such as the alkene or alkyne moieties used to link azacrown groups to the pyridyl ligands in complexes **1e** and **1f** (Fig. 1) [9, 10]. The UV–Vis absorption spectra of these

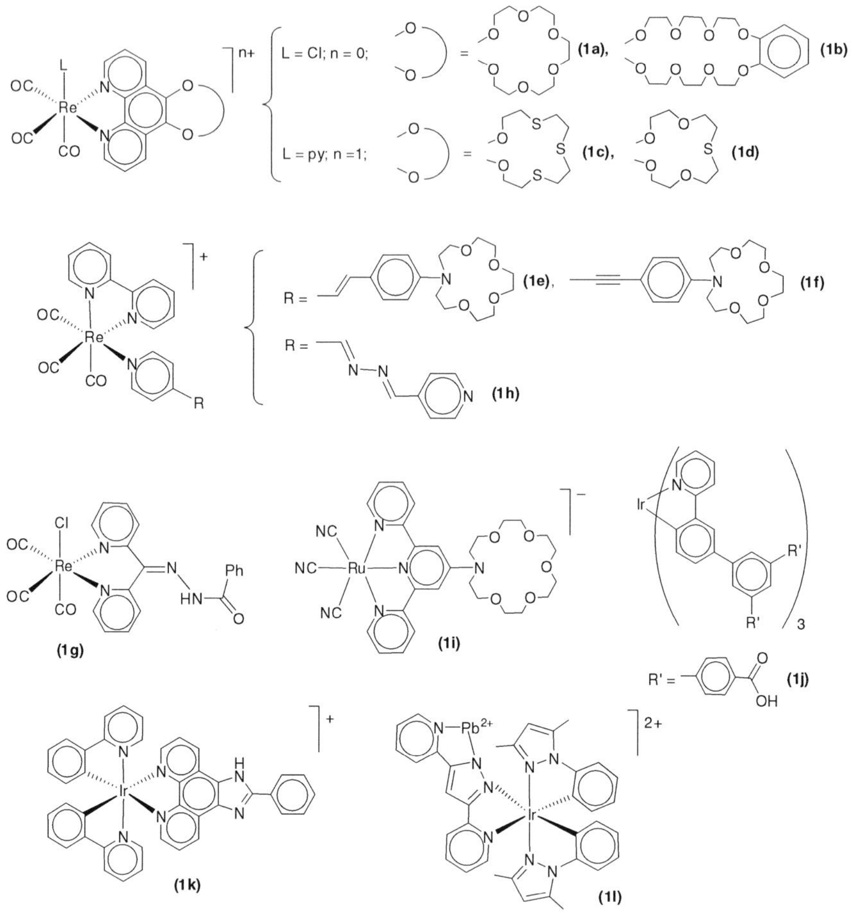

Fig. 1 Examples of Re, Ru and Ir complexes with pendant functionalities suitable for cation binding

compounds are dominated by strong intra-ligand charge transfer (ILCT) transitions localized on the pyridyl ligands (i.e. charge is transferred from the azacrown N atom to the ethenylpyridyl or ethynylpyridyl group) which hide the typical MLCT band. On protonation of the azacrown nitrogen, or complexation of Li^+, Na^+, K^+, Mg^{2+}, Ca^{2+} or Ba^{2+} into the macrocycle, the ILCT band shifts to lower wavelength with the magnitude of the shift being dependent on the cation. Recent work also include the development of pH-sensitive systems, such as *fac*-[Re(CO)$_3$(di-2-pyridylketonebenzoylhydrazone)Cl] (**1g**; Fig. 1), whose UV–Vis absorption spectrum has been shown to be very sensitive to the addition of acid or base, with acid concentrations as low as 10^{-9} M detected in DMF solution. This behaviour is attributed to the compound's ability to form strong H-bonding interactions with surrounding molecules through the amidic hydrogen atom [11]. The proton acceptor properties of the imine and/or pyridine N atoms in the ligand 4-pyridinecarboxaldehydeazine have also been exploited to prepare the pH-sensitive luminescent Re(I) complex **1h** (Fig. 1) as well as dinuclear Re^IRe^I and Re^IRu^{II} analogues [12].

2.2 Ruthenium(II) and Osmium(II) Complexes

The complex $[Ru(bpy)(CN)_4]^{2-}$ and its analogues are, like the Re-carbonyl species discussed above, typical MLCT $[d\pi(Ru) \rightarrow \pi^*(bpy)]$ chromophores that can be used as optical molecular sensors, usually exploiting the ability of the cyanide ligands to coordinate metal cations. For example, the interaction of $[Ru\{4,4'-di(^{tert}butyl)-2,2'-bpy\}(CN)_4]^{2-}$ with a variety of metal cations, such as Li^+, Na^+, K^+, Cs^+, Ba^{2+} and Zn^{2+}, in acetonitrile results in a blue-shift of the ^{1}MLCT absorption and the 'switching on' of strong ^{3}MLCT luminescence. The energy and intensity of the emission depends on the nature of the cation present, with λ_{em} varying from 646 nm (Cs^+) to 537 nm (Zn^{2+}) and the highest quantum yield (0.07) obtained for Ba^{2+}. Structural studies using the less soluble analogue $[Ru(bpy)(CN)_4]^{2-}$ showed that both 'end-on' and 'side-on' coordination of CN occurs, depending on the size and charge of the cation [13]. The rich luminescence properties of $[Ru(bpy)(CN)_4]^{2-}$ have also been used to design interesting logic gates responding to two chemical stimuli (acid and base), based on the self-assembly of the Ru(II) complex with a naphthyl-substituted dendritic host [14]. Addition of a pendant azacrown ligand to $[Ru(terpy)(CN)_3]^-$ has resulted in a new water-soluble Ru(II) chromophore (**1i**; Fig. 1) which exhibits selective cation-binding behaviour (through the azacrown pendant) and pH-sensitivity (through protonation of the cyanide ligands). The complex has been shown to be a good candidate for sensing subtle humidity changes and to serve as a mobile-phase additive in high-performance liquid chromatography (HPLC) for the separation of alkali and alkaline-earth metal cations, as well as amino acids [15]. The cation-binding properties of a Ru(II)-carbene derivative with a benzo-15-crown-5 pendant have also been reported [16].

Cyano-Os(II) complexes can also be used as cation sensors in a similar manner as their Ru analogues. Thus, $[OsL_2(CN)_2(N-N)]$ (L is PPh_3, PMe_3 or DMSO) containing a variety of phen and bpy (N–N) ligands have recently been shown to act as luminescent sensors towards Zn^{2+} through ion binding of the cyano ligands [17].

2.3 Iridium(III) Cyclometallated Complexes

There are numerous instances where the MLCT phosphorescence of cyclometallated Ir(III) complexes is exploited for sensing applications. In a recent example, the emission intensity of the water soluble Ir complex **1j** (Fig. 1) has been shown to increase with an increase in pH, in accord with the degree of ionization of the carboxylic acid in the ligand. Addition of metal cations, however, quenched the luminescence, with a more significant change found for divalent ions Mg^{2+}, Cu^{2+} and Hg^{2+}, compared to monovalent alkali cations [18]. Derivatives containing imidazo-phenanthroline units, such as **1k** (Fig. 1) are also sensitive to the addition of H^+, possibly through interactions with the imidazolyl group, although these compounds have been mainly applied in anion sensing [19]. In another recent example (**1l**; Fig. 1), the cation-binding 3,5-di(pyridyl)pyrazole unit was used to preferentially bind Pb^{2+} vs Hg^{2+}, Ca^{2+} or alkali cations. The sensing of Pb^{2+} is based on the significant quenching of the room-temperature phosphorescence upon forming a 1:1 adduct with the Ir(III) complex, and can be performed in acetonitrile solution or by supporting the Ir(III) compound on a solid which can then be put in contact with an aqueous solution of Pb^{2+}. The latter provides the basis for a convenient detection device [20].

2.4 Ferrocene-Based Sensors

Although more often used as electrochemical sensors, ferrocene (Fc) derivatives can also act as optical sensors because their UV/Vis absorption properties can be readily modified upon binding of guests near the Fc moiety. This is usually achieved through perturbation of the characteristic low energy MLCT transition of ferrocene (at ca. 400–500 nm). Often dual sensors with combined electrochemical and optical responses are prepared, although only changes in the optical properties of the compounds will be discussed herein. Fc sensors are also particularly versatile because it is relatively easy to introduce a large variety of binding groups and functionalities on the cyclopentadienyl (Cp) ligands. For example, the doubly functionalized colourless Fc derivative **2a** (Fig. 2) can bind F^- through its urea fragment, and K^+ via the crown ether. Addition of F^- produces a yellow compound in acetonitrile, but the colour is quenched by the subsequent addition of K^+. This 'on–off' switching effect sets the basis for a logic gate [21]. Basurto et al. [22] have prepared a family of multi-functionalized Fc derivatives (e.g. **2b** and **2c**, Fig. 2) showing significant spectral changes upon addition of a variety of metal ions (e.g. Cu^{2+}, Hg^{2+}, Zn^{2+}, Cd^{2+}, Pb^{2+}, Fe^{3+}, Al^{3+} and Ag^+) as well as several anions, including the selective naked-eye colorimetric detection of Cu^{2+}. Fc ligands with diamino–diimido functionalities have also been proposed for the combined potentiometric/spectrophotometric detection of Cu^{2+} [23]. Diferrocenyl derivative **2d** (Fig. 2) allows naked-eye detection of Mg^{2+} and shows selective UV/Vis spectral changes towards this cation, while not responding to Ca^{2+} or alkaline metal ions. The metal-sensor interaction is believed to occur through the C=N–C group with no participation of the OH group [24]. Analogous Mg^{2+} detection properties have been also described for the related pyridine derivative **2e** [25], and the triferrocene complex **2f** [26], whereas Fc-ruthenocene derivatives (e.g. **2g**) act as selective

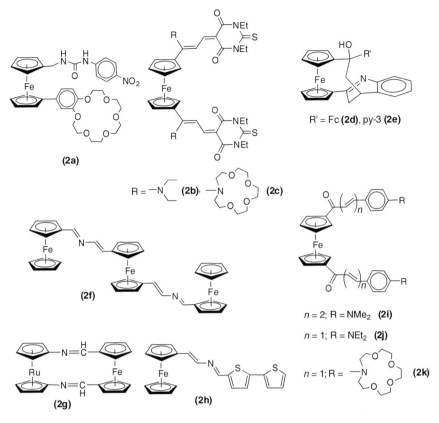

Fig. 2 Examples of ferrocene (Fc) derivatives used for cation sensing

naked-eye sensors towards Zn^{2+} in preference to many other mono- and di-valent cations (Fig. 2) [27]. In the two latter complexes, metal coordination is assumed to occur mainly through the aza-N atoms. The introduction of S-donor groups in this type of ligands (e.g. **2h**, Fig. 2) allows the selective detection of soft metals, such as Cd^{2+}, Hg^{2+} and Pb^{2+} [28, 29]. A series of other Fc-ruthenocene derivatives with interesting cation-sensing properties have also been reported [30, 31].

Despite the fact that Fc units normally act as fluorescence quenchers, it is possible to prepare fluorescent Fc derivatives that respond to the presence of guests. Thus, it has been shown that the absorption and emission spectra of compounds **2i–2k** (Fig. 2) change significantly in the presence of Ca^{2+} or Ba^{2+} but remain largely unchanged on addition of Li^+, Na^+ or K^+, with the main metal binding sites located at the carbonyl group (**2i**, **2j**) or azacrown (**2k**) groups. The detection ability of the compounds is, however, limited by the fact that the fluorescence intensity does not change monotonically with the concentration of the metal ion, which is attributed to the formation of various species with different stoichiometries in solution [32–34]. Analogous Fc derivatives with only one arm are not fluorescent but show UV–Vis spectral changes on addition of Ca^{2+} [35]. Ferrocene derivatives containing anthracene groups also act as cation sensors through enhancement of the structured fluorescent band of the anthracene moiety on metal complexation [36, 37].

2.5 Platinum(II) Complexes

In recent years, Pt(II) alkynyl terpyridine complexes have been shown to be versatile ion sensors [6]. For example, complexes with amino–alkynyl ligands (**3a**; Fig. 3) act as pH sensors, derivatives with crown ether pendants (**3b**; Fig. 3) have been used as sensors for mono- and di-valent cations, such as Li^+, Na^+, Mg^{2+}, Ca^{2+}, Cd^{2+} or Zn^{2+}, and analogous compounds containing flavones as pendant ligands have shown selective binding towards Pb^{2+} [38–42]. The origins of the changes observed in the absorption and emission spectra vary, with some systems showing switching between different excited states in the Pt(II) complex on ion binding. The related dinuclear calix[4]crown complex (**3c**; Fig. 3) also shows marked luminescence enhancement on cation binding, with a more selective response towards K^+ [43]. The supramolecular Pt rectangle (**3d**; Fig. 3) has been proposed as an optical sensor for Ni^{2+}, Cd^{2+} and Cr^{3+} in solution. The UV–Vis spectrum of the rectangle shows

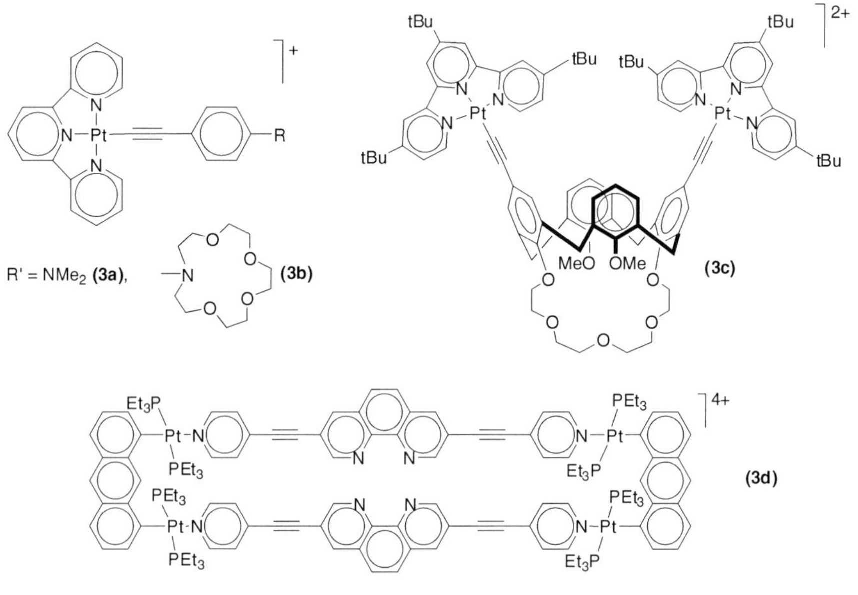

Fig. 3 Examples of Pt(II) complexes used for cation sensing

significant changes in the presence of the cations, which coordinate to the phenanthroline units without disturbing the supramolecular structure [44].

2.6 Gold(I) Complexes

The tendency of Au(I) and other d^{10} metals to form metal–metal interactions is often at the origin of intense emission, which has been exploited to prepare cation sensors; e.g. the emission can be switched 'on' by favouring the formation of metal–metal contacts upon ion coordination. This approach has been successfully applied by Yam et al., who used dinuclear Au(I) complexes containing ligands, such as calix[4]crown, crown ether or alkynyls, capable of trapping an additional metal ion and concomitantly altering the emission of the compounds by shortening significantly the Au–Au distance (e.g. **4a** and **4b**; Fig. 4) [45–48]. Recent work in related digold(I) complexes with alkynyl binding units has shown that cation binding can lead to complicated dimeric structures [49, 50]. Interestingly, Yam et al. have combined the binding and optical properties of Au(I)-alkynyls with the photoisomerization of azobenzene moieties in a macrocyclic complex, the conformation

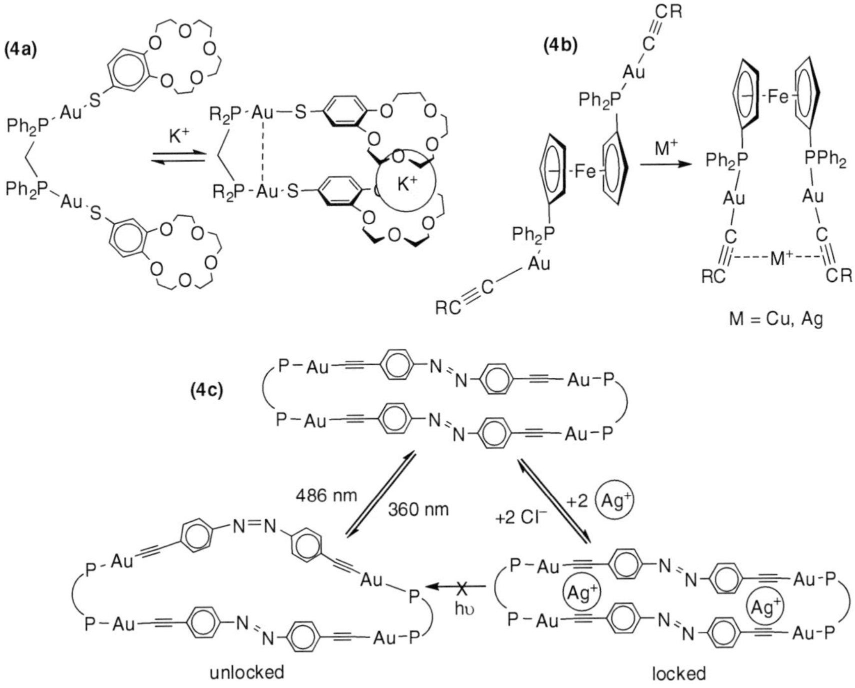

Fig. 4 Examples of applications of Au(I) complexes in optical sensing

of which can be controlled by addition or removal of Ag(I), thus setting the basis for a dual-input molecular logic photoswitch (**4c**, Fig. 4) [51].

3 Anion Sensors

3.1 Rhenium(I) Carbonyl Complexes

The Re(I) tricarbonyl moiety, in combination with pyridyl ligands, has led to a number of both neutral and charged species [52]. Sloan et al. initially isolated a self-assembled Re(I)–Pd(II) square (**5a**; Fig. 5) which demonstrated an increase in emission with the inclusion of ClO_4^- within the cavity and a binding constant of

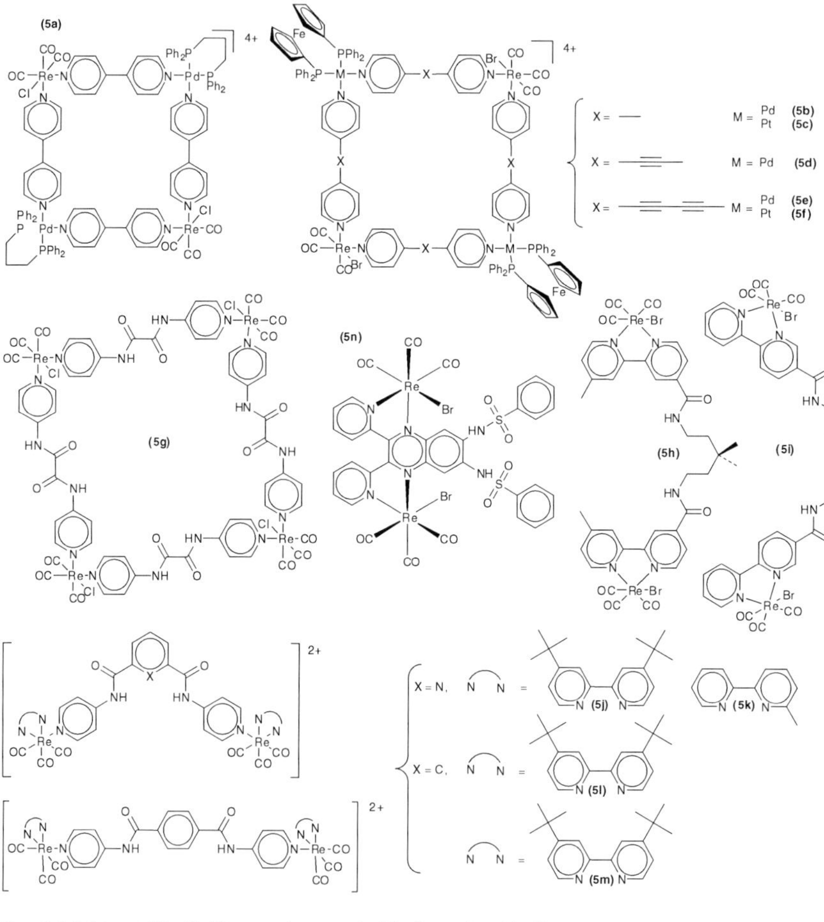

Fig. 5 Multimetallic Re(I) complexes suitable for anion binding

660 M^{-1} [53]. The much larger and rigid cavities (**5b–5f**; Fig. 5) from the Lees group showed an unusual change in their luminescence on the introduction of BF$_4^-$ and PF$_6^-$ anions, initially with a dramatic decrease in the observed fluorescence, followed by a significant increase. It is noted that there was no observed change with ClO$_4^-$, OAc$^-$ and OTf$^-$ [54]. More recently the neutral molecular square (**5g**) reported by Tzeng et al. has shown colorimetric changes on the introduction of a range of inorganic anions with selectivity for F$^-$ over CN$^-$ and OAc$^-$, and no perturbation observed for Br$^-$, PF$_6^-$, BF$_4^-$, ClO$_4^-$, NO$_3^-$ or HSO$_4^-$ [55].

The neutral dimetallic complex (**5h**; Fig. 5) reported by Beer et al., which demonstrated reasonable affinities for H$_2$PO$_4^-$ in ^{1}H NMR titration studies [56], inspired a recent study from Pelleteret et al. They have recently reported a series of neutral bimetallic Re(I) complexes (**5i**; Fig. 5), bridged by a flexible ethylene glycol diamide chain [57, 58]. This complex shows a significant increase in fluorescence upon the introduction of H$_2$PO$_4^-$ in non-protic solvents, which appears to arise from the removal of intra-molecular hydrogen bonds between the metal carbonyl and the free amide by the competitive oxo-anion. A series of structurally similar dinuclear luminescent Re(I) tricarbonyl anion receptors, featuring amide-type anion binding sites (**5j–5m**; Fig. 5) have also been shown to display outstanding sensitivity and selectivity toward a variety of anionic species [59, 60]. The intensity of these highly emissive positively charged species was significantly quenched by as much as 10%, even in the presence of only 10^{-8} M CN$^-$ or F$^-$ anions. The bis (sulfonamide) complex **5n** (Fig. 5) also exhibits considerable colorimetric change on the introduction of F$^-$, CN$^-$ and OAc$^-$, attributed to a de-protonation which is exacerbated by the Lewis acidity of the metal and can be used to modulate the amidic pK_a [61].

Building on the ability of pendant cation receptors described above, the emission intensity of Re(I) complexes with pendant crown-ether groups (**6a–6c**; Fig. 6) was shown to increase on the addition of OAc$^-$ with a concomitant 5 nm hypsochromic shift. The observed stability constants were larger when a K$^+$ ion sits within the macrocycle [62]. Similarly, the calix[4]arene appended complex **6d** (Fig. 6) demonstrates a selectivity for OAc$^-$ with a marked revival of the quenched emissive behaviour on the introduction of the anions [63]. In another example, the inclusion of the positively charged pseudorotaxane thread within a macrocyclic complex (**6e–6h**; Fig. 6), assembled around a Cl$^-$ ion, resulted in a significant enhancement in the fluorescence intensity [64]. This may be as a consequence of the increasing rigidity of the complex upon the interaction with the anion. This was further exemplified with a stoppered rotaxane thread (**6i**, Fig. 6), where removal of the templating anion, and subsequent replacement with Cl$^-$, NO$_3^-$ and, significantly, HSO$_4^-$ in acetone gave a similar increase in the quantum yield [65]. A couple of charged Re(I) complexes (**6j** and **6k**; Fig. 6) have been shown by Lo et al. to recognize F$^-$, OAc$^-$ and H$_2$PO$_4^-$ with pK_s values ranging from 3.53 to 4.94 in acetonitrile [66]. While the anion binding properties of **6k** can be reflected by changes in the fluorescence, complex **6j** shows perturbations in both the absorption and emission spectra.

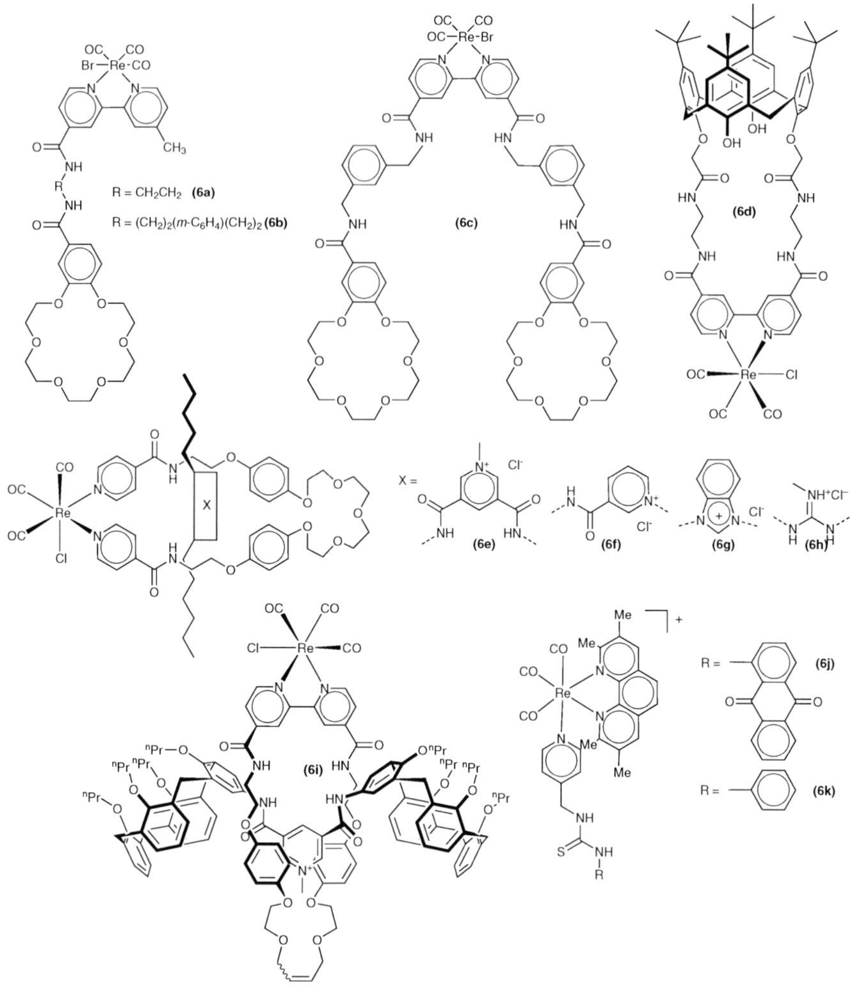

Fig. 6 Re(I) complexes with appended binding domains for anion recognition

3.2 *Platinum(II), Ruthenium(II) and Iridium(III) Complexes*

In recent years, several other organometallic fluorogenic centres have been considered for inclusion in anion sensing materials. Two Pt(II) terpyridyl alkynyl complexes (**7a** and **7b**, Fig. 7) show a dramatic colorimetric response to F^-, OAc^- and $H_2PO_4^-$, arising from interaction of the phenolic proton with the electronegative anion, and perturbation of the ligand-to-ligand charge transfer. Complex **7b** shows selectivity for F^- due to complete de-protonation [67]. In addition, complex **7a** has a considerable quenching in fluorescence on the addition of these anions. Similarly, a series of Ru(II) alkynyl complexes (**7c**–**7g**; Fig. 7) show a dramatic

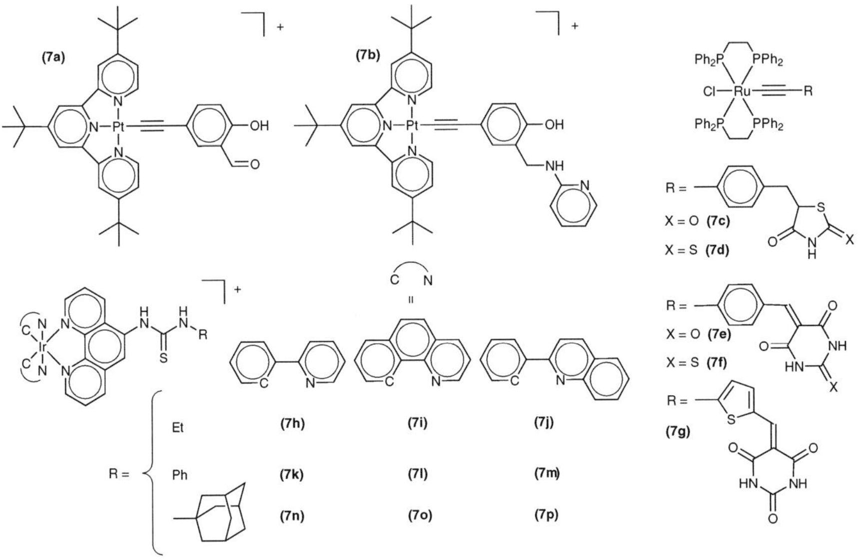

Chromo- and Fluorogenic Organometallic Sensors 155

Fig. 7 Pt, Ru and Ir complexes with anion recognition behaviour

colour change, visible to the naked eye, on the addition of a range of anions including OAc⁻, $H_2PO_4^-$, HSO_4^-, Cl⁻ and Br⁻, although selectivity for F⁻ is evident, and there is a dependence on the electron density on the appended functional group [68]. Bucking this trend, a series of luminescent cyclometallated Ir(III) poly-pyridine thiourea complexes (**7h–7p**; Fig. 7), have shown selectivity for OAc⁻ over both F⁻ and $H_2PO_4^-$ [69]. The thiourea moieties of the complexes permit a 1:1 stoichiometry with the anion, as demonstrated by a significant quenching of the emissive ^{3}MLCT state. This behaviour appears to be independent of the appended functional groups.

3.3 Ferrocene-Based Sensors

It has already been noted that Fc complex **2a** gives a colorimetric change in the presence of F⁻ [21]. In addition to the electrochemical response [70, 71], the attachment of suitable chromophores/fluorophores to Fc has permitted optical responses to anions. This was initially demonstrated by Beer et al., who showed that the inclusion of $H_2PO_4^-$ in a cleft between a Ru(II) and two Fc groups (**8a**; Fig. 8) induced a significant increase and a 30 nm red-shift in emissive behaviour [72]. In the structurally similar Fc and cobaltocinium complexes (**8b** and **8c**; Fig. 8), the rate of the quenching energy transfer between the two metal centres could be perturbed by the inclusion of Cl⁻ in the intra-metallic cavity, resulting in a significant switching of the Ru MLCT based fluorescence [73]. The

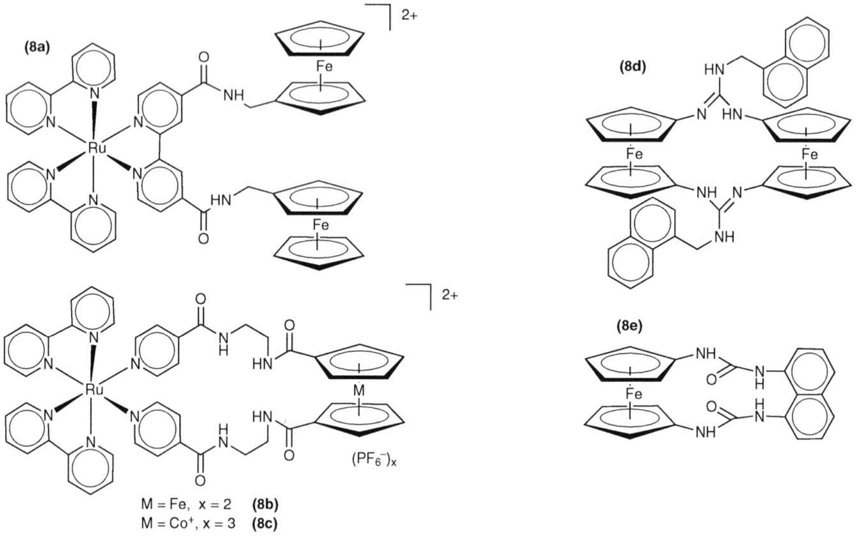

Fig. 8 Examples of ferrocene (Fc) derivatives used for anion sensing

guanidinoferrocene receptor (**8d**; Fig. 8), in addition to potentiometric detection of F^-, OAc^-, HSO_4^- and $H_2PO_4^-$, can act as a fluorescent chemosensor for Zn^{2+}, Ni^{2+} and Cd^{2+} metal ions. Furthermore, proton induced complexation provides a versatile means of sensing selectively NO_3^- via fluorescence quenching [74]. A urea centred complex (**8e**; Fig. 8) with a weak naphthalene emission when excited at 310 nm undergoes a 13-fold enhancement on the introduction of F^-, and a weaker, but significant response to $H_2PO_4^-$ [75].

4 Small Molecule Detection

4.1 Detection of Gases and Volatile Organic Compounds

The optical detection of vapours or gases by metal complexes often requires the sensor to show a response in the solid-state (i.e. detection takes place either by crystals of the complex, or by the compound supported on a solid matrix, cast as a film, etc.). Changes of the optical properties of the sensor upon absorption of the analyte are related to (1) quenching of the sensor's luminescence (e.g. by O_2), (2) disruption/modification of non-covalent interactions in the crystal packing of the sensor (e.g. H-bonding, π–π stacking, metal–metal contacts), and/or (3) change of the coordination environment of the metal centre (e.g. through coordination of the analyte to the metal). Although a porous sensor structure is not a pre-requisite, a relatively 'open' and flexible crystal packing favours the inclusion of the volatile and its interaction with the sensor.

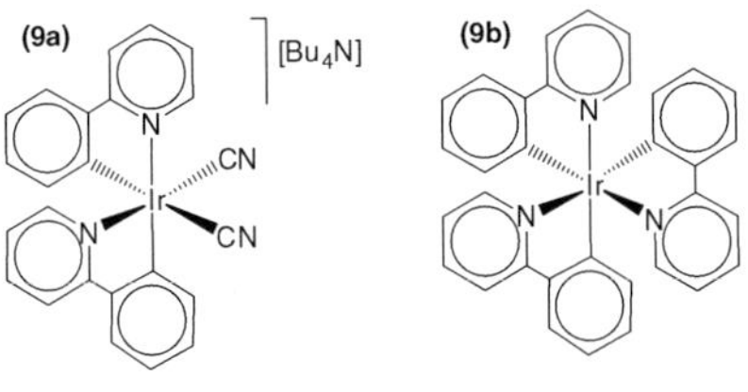

Fig. 9 Ir(III) complexes used for O_2 sensing

Detection of dioxygen is often based on the quenching of the ^{3}MLCT emission of d^6-metal complexes. In a recent example, cyclometallated compounds [Bu$_4$N][Ir(ppy)$_2$(CN)$_2$] (**9a**) and [Ir(ppy)$_3$] (**9b**; Fig. 9) dissolved in poly((n-butylamino)thionylphosphazene) polymer films have been shown to have high sensitivity to oxygen quenching [76]. Detection of SO$_2$ has been achieved using Pt(II) complexes with N,C,N'-pincer ligands which, in their crystalline state, are able to coordinate reversibly the gas giving rise to a colour change. Based on this process, multi-metallic dendritic sensors were also developed [77]. Coordination of ethylene to Pd(II) or Pt(II) square planar complexes has recently been applied to prepare a colorimetric ethylene detector based on a 2,9-di-n-butylphenanthroline Pd(II) complex supported on silica [78].

The optical properties of Pt(II) complexes are often associated with the formation of Pt^{II}–Pt^{II} stacked structures [79]. Thus, the well-known intense colours and vapochromic properties of complexes [Pt(CNR)$_4$][M(CN)$_4$] (M is Pt, Pd; R is alkyl or aryl) are attributed to the presence of extended metal–metal chains of alternating anions and cations [80, 81]. The inclusion of vapour guests by the solids gives rise to colour changes (and/or changes in emission) which are mainly related to disruption of the metal–metal interactions, but also to changes in other non-covalent contacts throughout the structure. In recent years new analogues have been prepared and applied in novel sensing devices [82, 83], including humidity sensors [84]. Mann et al. [85] have shown that thermal re-arrangement of [Pt(p-CN–C$_6$H$_4$–C$_2$H$_5$)$_4$][Pt(CN)$_4$] also yields the vapochromic isomer cis-[Pt(p-CN–C$_6$H$_4$–C$_2$H$_5$)$_2$(CN)$_2$], whose structure also consists of linear Pt^{II}–Pt^{II} chains. Exposure of the crystalline solid to aromatic volatile organic compounds (VOCs), such as toluene, benzene, chlorobenzene, p-xylene or mesitylene, produces a reversible blue-shift in the emission. It is noted that the *trans* isomer does not exhibit vapochromic behaviour, which is attributed to its less open structure, with shorter metal–metal distances, and an efficient π–π stacking of isocyanide ligands [86]. Rod-like crystals over 500 μm in length have been obtained for the analogue cis-[Pt(CN^{t}Bu)$_2$(CN)$_2$] using an injection–reprecipitation method. These Pt^{II}–Pt^{II} stacked 'wires' exhibit intense green emission which shifts significantly, and/or increases in intensity, on exposure to VOCs [87]. [Pt(CN)$_2$(4,4'-dicarboxy-2,2'-bpy)] forms several intensely coloured polymorphs depending on pH and crystallization solvents. Interconversion between the various polymorphs occurs in solution and also on exposure of the crystals to different solvent vapours. It is suggested that bulky substituents on the bipyridine ligand favour the formation of cavities in the crystal lattices and the subsequent inclusion of guests [88]. The red potassium salt [K(H$_2$O)][Pt(bzq)(CN)$_2$] (bzq is 7,8-benzoquinolinato) has been used to prepare vapochromic films that become yellow in the presence of solvent vapours, such as dichloromethane, methanol, ethanol, acetone, tetrahydrofuran, or acetonitrile, after exposure times ranging from 5 s (methanol) to 45 min (tetrahydrofuran) [89]. In other examples, Pt–Pt interactions are not directly involved in the vapochromic behaviour of the complexes [90, 91]. For example, crystals of the intensely luminescent solvated complex **10a** $6CHCl_3 \cdot C_5H_{12}$ (Fig. 10) change colour on desolvation and show a significant decrease in emission intensity [91]. The original colour and intense emission are restored upon exposition to the vapour of halogenated solvents or small polar VOCs (e.g. THF, acetone, or diethyl ether), but not to aromatics, methanol or ethanol. It has been shown that the solvated structure contains voids or solvent channels with an extended network of CH–π and π–π interactions, which rearrange on desolvation, giving a more compact packing, and presumably favouring quenching of the luminescence via inter-molecular dipole–dipole interactions. The incorporation of halogenated and small polar solvents is favoured by the formation of relatively strong CH–π, π–π, CH–X and X–X interactions, whereas the size of the voids may be too small to host aromatic VOCs. In the case of derivatives **10b** and **10c** (Fig. 10) [90] the alkynyl moieties act as convenient receptors for chlorinated solvents via multiple CH–π(C≡C) interactions.

Since the early example of Balch et al. [92] of solvent-induced luminescence of a trigold(I) complex, attributed to the presence of columnar Au^I–Au^I interactions, other examples have been proposed as efficient VOC detectors based on extended metal–metal contacts involving Au(I), as well as other d^{10} metal ions. For example, Fernandez et al. described the vapochromic and vapoluminescent behaviour of {Tl[Au(C$_6$Cl$_5$)$_2$]}$_n$ (**10d**, Fig. 10), which exhibits a 3D network of Au^I–Tl^I interactions

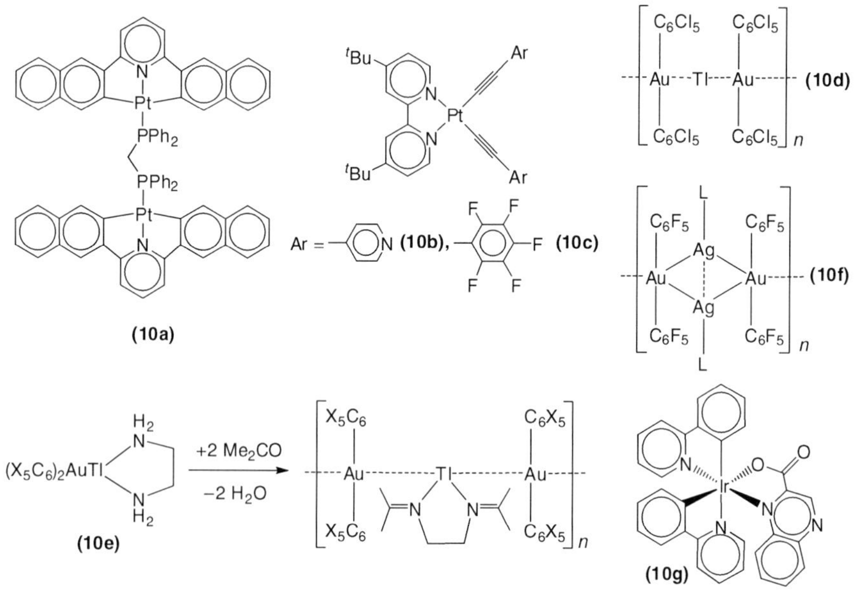

Fig. 10 Examples of vapochromic/vapoluminescent metal complexes

with channels into which the VOCs can diffuse [93, 94]. Reversible changes of colour are observed when the solid is exposed to a variety of VOCs, such as acetone, acetonitrile, triethylamine, acetylacetone, tetrahydrothiophene, 2-fluoropyridine, tetrahydrofuran, and pyridine vapours. The luminescence of **10d** has its origin largely in the Au^I–Tl^I contacts, and binding of the Tl atoms to the VOCs are believed to be responsible for the observed behaviour. Interestingly, the colour change observed when $[(X_5C_6)_2AuTl(ethylenediamine)]_n$ (X is Cl, F; **10e**) reacts with ketone vapours in the solid-state is associated to condensation reactions at the amine moieties, promoted by their coordination to Tl (Fig. 10) [95]. Polymeric Au^I/Ag^I complexes $\{Ag_2L_2[Au(C_6F_5)_2]_2\}_n$ (L is Et_2O, Me_2CO, THF, CH_3CN; **10f**, Fig. 10) also show fast vapochromic and vapoluminescent behaviour towards a variety of VOCs. Recent studies have proven that substitution reactions between the ligand at the Ag atoms and the VOCs are responsible for the observed behaviour, and that the emission is localized in the tetranuclear Au^I/Ag^I cores [96]. This type of Au^I/Ag^I polymers can be incorporated into fibre optics for optical measurements [97–102]. The coordination polymers $\{Cu[Au(CN)_2](solvent)_2\}_n$, which can be obtained in two polymorphic forms depending on the solvent, exhibit visible colour changes on exposure to various volatiles, including water, MeCN, DMF, dioxane, pyridine and NH_3. This is due to their coordination to the Cu(II) centres, each analyte modifying the crystal field splitting differently [103]. A related compound, $\{Zn[Au(CN)_2]\}_n$, responds to NH_3 with detection limits as low as 1 ppb [104].

The Ir(III) cyclometallated complex **10g** (Fig. 10) shows selective vapochromic and vapoluminescent properties in response to acetonitrile or propiononitrile vapour, while remaining unchanged for other VOCs [105]. Crystals of the compound change from a black to a red polymorph in less than 1 min on exposure to the solvent vapours; at the same time, strong emission is switched 'on' (red form). Structural analyses show that the black form exhibits shorter π–π stacking distances between the quinoxaline ligands of neighbouring molecules, which favours dipole–dipole interactions and quenching of the luminescence.

4.2 Detection of Small Organic Molecules in Solution

The detection of small organic species in solution has proven to be a significant challenge, and has received surprisingly little attention. The *fac*-carbonyl Re molecular squares, bearing organometallic fragments, have featured in a number of studies, building on their observed recognition of both anionic and volatile organic species [52]. The large Re(I) cornered square bridged by zinc porphyrin moieties **11a** (Fig. 11), demonstrate considerable red shifts in the porphyrin based fluorescence on the addition of millimolar quantities of pyridine, but not with toluene [106]. In the simple neutral cyclophanes **11b–11d** (Fig. 11), which posses a variety of different rigid spacers, the luminescence quenching rate constants, k_q, of the ^{3}MLCT excited state have been investigated in the presence of a variety of

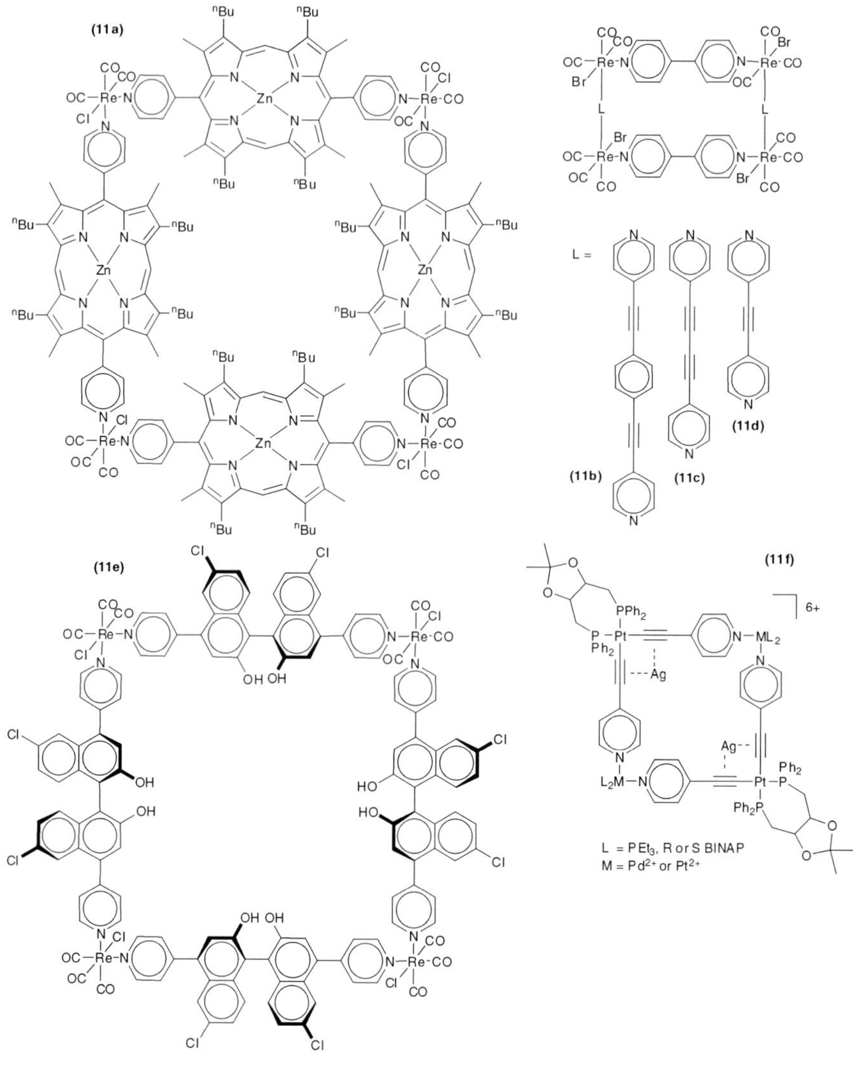

Fig. 11 Examples of optical detectors for small molecules

aromatic amines such aniline and were found to be higher than those for simple monomeric Re(I) complexes [107]. In another example an enantiopure square **11e** (Fig. 11) undergoes luminescence quenching with the chiral alcohol, R- or S-2-amino-1-propanol, and in the process demonstrates an enantioselectivity of 1.22, determined by differential quenching rates [108]. The optically active mixed, neutral Pt/Pt/Ag and Pt/Pd/Ag macrocyclic complexes (**11f**; Fig. 11) have undergone an exploration for the inclusion of several diamines demonstrating a significant interaction with the neutral guests tetramethylpyrazine and phenazine, detected by perturbations in the circular dichroism spectra [109].

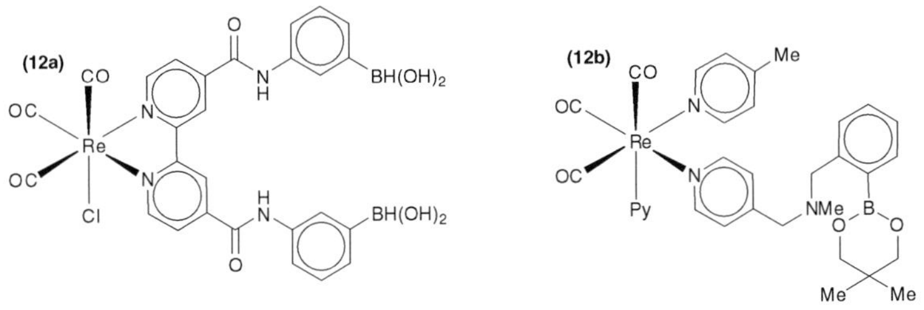

Fig. 12 Examples of Re(I) complexes for the optical detection of glucose

The detection of sugar has attracted a number of studies. The neutral diboronic acid complex **12a** (Fig. 12) was reported by Yam et al., illustrating a preference for mono-saccharides and a marginal preference for D-fructose ($pK_s = 2.40$ in DMSO), through a perturbation in the ^{3}MLCT transition, and an increase in the observed emission [110]. There is, however, considerable complexity to the system in aqueous solution given the pH dependence of the boronic acid groups [111]. Glucose testing with a structurally similar complex **12b** (Fig. 12) showed a significant dependence on the solvent system used; in methanol, a 55% fluorescence intensity increase was observed with a change in glucose concentration from 0 to 400 mg dL^{-1}, although in a methanol-phosphate buffered saline solution, no significant response was found to glucose at physiological pH [112].

A competitive indicator displacement approach has been successfully employed for sulfhydryl amino acids and short chain peptides using heterobimetallic donor–acceptor complexes: cis-[ML$_2$(μ-CN)$_2${Pt(DMSO)Cl$_2$}$_2$] (M is Fe^{2+}, Ru^{2+} and Os^{2+} and L is 2,2'-bpy). The cis-[ML$_2$(CN)$_2$] units are used as signalling indicators, and {Pt(DMSO)Cl$_2$} as both an acceptor group, and the receptor for the analytes. All three ensembles are able to produce specific colorimetric/fluorometric responses to cysteine, homocysteine and methionine, as well as the sulfhydryl-containing small peptide glutathione [113, 114]. In a very recent paper, an alternative strategy, employing an Ir(III) complex [Ir(pba)$_2$(acac)] (Hpba is 4-(2-pyridyl)benzaldehyde and acac is acetylacetone) has been shown to posses selectivity for homocysteine. Upon addition of the amino acid, in a semi-aqueous solution, a colour change from orange to yellow and a luminescent variation from deep red to green were evident to the naked eye, attributed to the formation of a thiazinane group from the free aldehyde group [115].

5 Sensing of Large Biomolecules

5.1 Detection of DNA

The development of probes for large biomolecules has attracted considerable attention using luminescent late transition metals, particularly Ru(II) polypyridyl

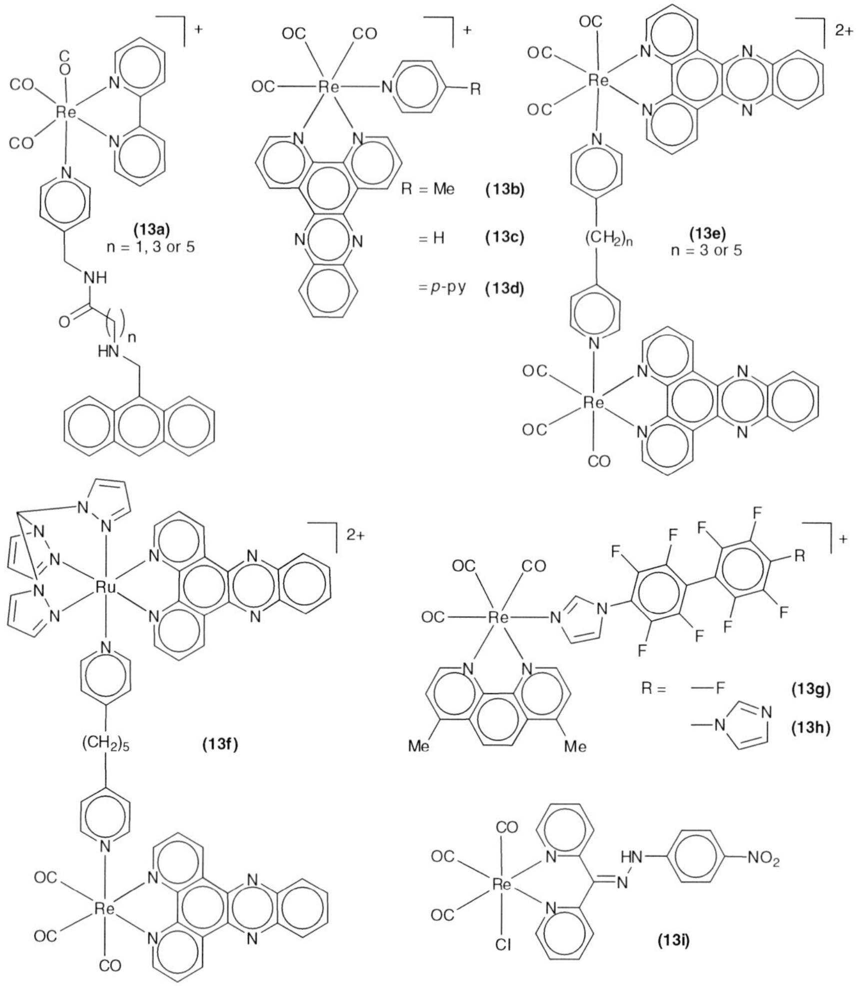

Fig. 13 Examples of Re(I) complexes for the optical detection of biomolecules

lumophores. Attention has been given to sensing the presence of DNA, and has been the subject of several review articles [116–118]. As with other areas of molecular detection, the fluorescent Re(I) moiety has been the subject of a number of studies. Schanze and Thornton isolated a series of complexes (**13a**, Fig. 13), structurally related to the anion recognition unit to **6j** (Fig. 6), bearing an anthracene unit [119, 120]. The complexes exhibit dual emission from both the metal ^{3}MLCT and the anthracene unit, although significant quenching is observed. On addition of DNA, hypsochromism of the anthracene absorption is detected, but not in the Re centred transitions, suggesting that the planar anthracene group is intercalating in the DNA. A dramatic increase was observed in the ^{3}MLCT emission in keeping with many transition based metal complexes associated with the hydrophobic DNA environment. Building on these results, a similar 'light-switch' interaction was observed with the Re(I) complex **13b** (Fig. 13), bearing the intercalating ligand dipyrido[3,2-*a*:2′,3′-*c*]phenazine (dppz). In aqueous solution the complex is non-luminescent, whereas ^{3}IL$_{dppz}$ phosphorescence is observed from the aqueous DNA bound complex [121]. Similar behaviour was observed with two structurally related complexes (**13c** and **13d**, Fig. 13), which exhibited potential for photo-cleavage of DNA [122, 123]. Recent developments include the consideration of bimetallic complexes, such as **13e**, where a hypsochromic shift in both the MLCT and IL transitions when added to buffered DNA [124]. Studies with the hetero-dinuclear complex **13f** reveal energy transfer between the Re(I) and the Ru(II) centres, and a light-switch behaviour in the emission on the introduction of DNA [125].

5.2 Detection of Proteins

The use of fluorescent organometallic complexes to label biological substrates is beginning to provide some exciting alternatives to the more traditional organic dyes [126], with suitable iridium [127–129], rhodium [130], platinum [131], rhenium [132–135] and osmium [136] examples having recently been reported. The Re diimine 'wires' (**13g** and **13h**, Fig. 13), have been shown to form complexes with the nitric oxide synthase mutant δ114 [137]. Steady-state luminescence measurements with **13h** establish a dissociation constant of 100 nM, while **13h** binds with a K_d of 5 μM, causing partial displacement of water from the haeme iron. The optical behaviour of (**13i**, Fig. 13), in the presence of L-methionine, and chemotactic *N*-formyl-amino acids: *N*-formyl-L-methionine, *N*-formyl-L-glycine and *N*-formyl-L-phenylalanine shows the complex to be insensitive to L-methionine, but highly sensitive to the *N*-formylamino acids at concentrations of less than 10^{-5} M in non-aqueous polar solvents, and is dependant on the polarity of the side chain of the amino acids.

Considerable progress has been made by the group of Lo in developing organometallic probes capable of recognizing the presence of specific classes of proteins, and has recently been the subject of a detailed review [138]. To highlight some recent advances, a large number of *fac*-carbonyl Re(I) (**14a–14c**, Fig. 14) [139–141] and cyclometallated Ir(III) complexes (**15a–15c**, Fig. 15) [142–144] bearing biotin moieties have been reported. Each of these complexes exhibit the characteristic ^{3}MLCT emission in solution, which was significantly enhanced in the presence of the glycoprotein avidin. Furthermore, the inclusion of the extended planar ligands dppz and dppn has been shown to permit similar enhancement to be observed in the presence of DNA [141, 145].

In another exciting development, the inclusion of a thiocyanate group into the ligands of complex **14b** has allowed the complex not only to bind to avidin, but also to be tagged to bovine serum albumin (BSA) [146]. The inclusion of indole groups

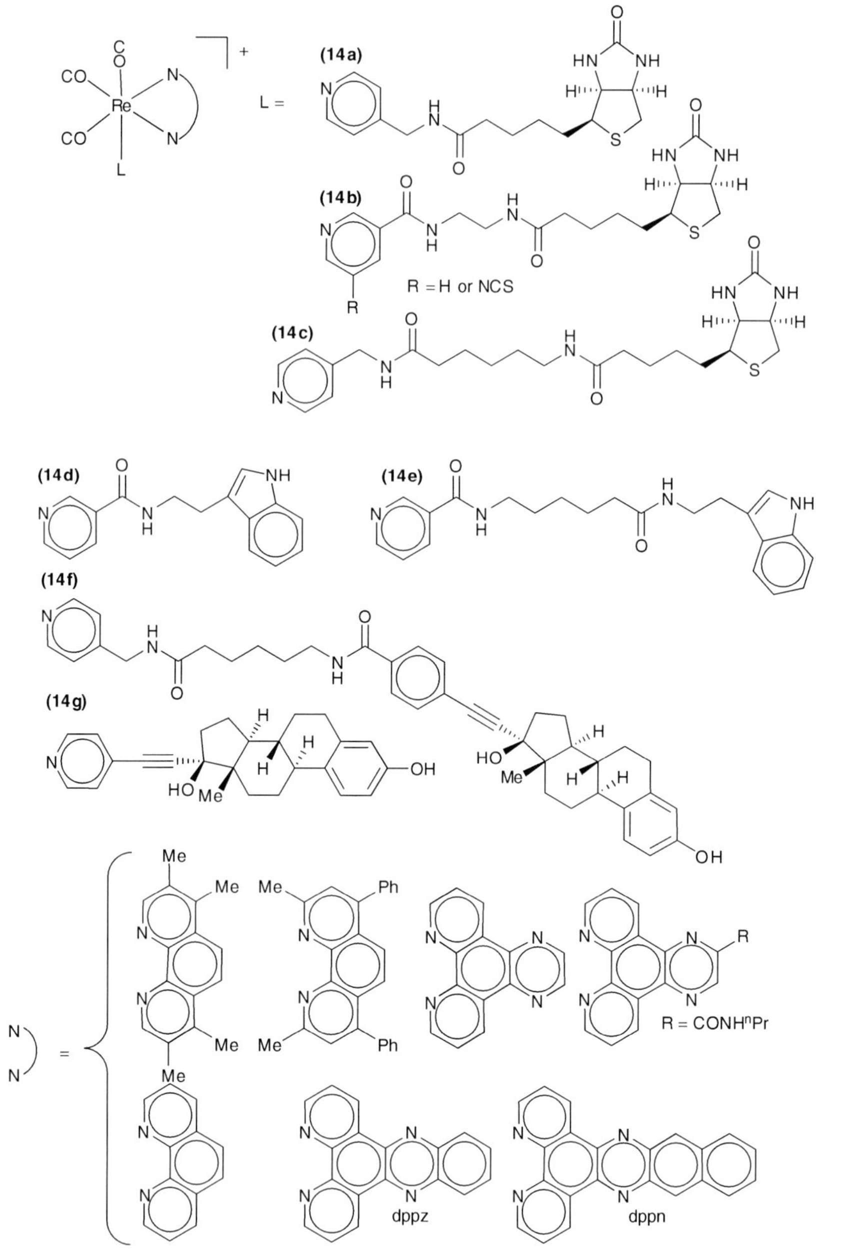

Fig. 14 Examples of Re(I) complexes used in the optical detection of biomolecules

into both the Re(I) (**14d** and **14e**, Fig. 14) chromophores [147, 148] show enhanced emission in the presence of indole binding proteins including BSA and tryptophanase, despite having excellent emission in aqueous solution. Similarly, the estrodiol

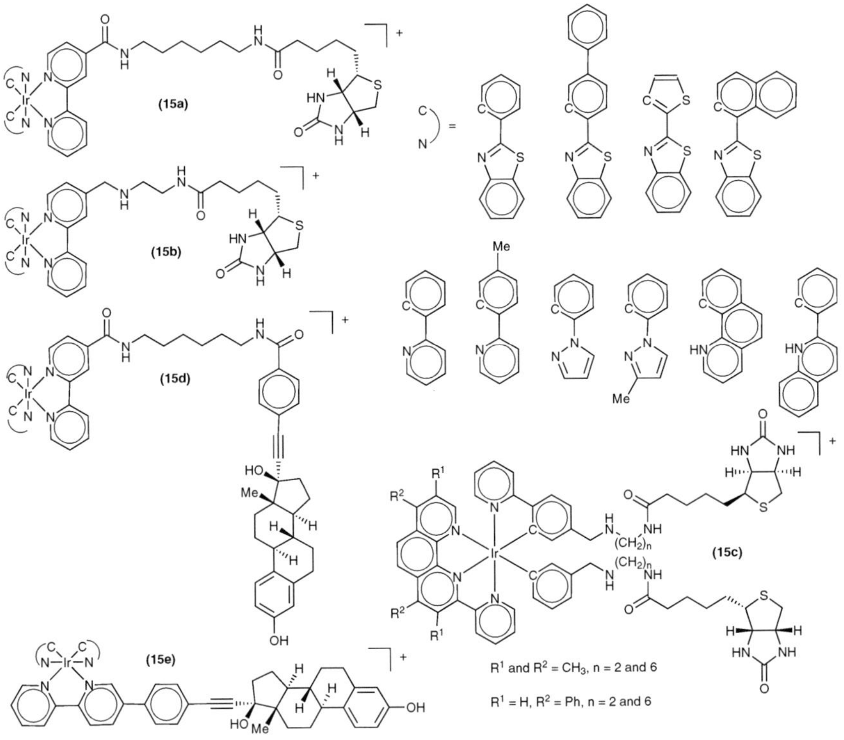

Chromo- and Fluorogenic Organometallic Sensors

Fig. 15 Examples of Ir(III) complexes for the optical detection of protein interactions

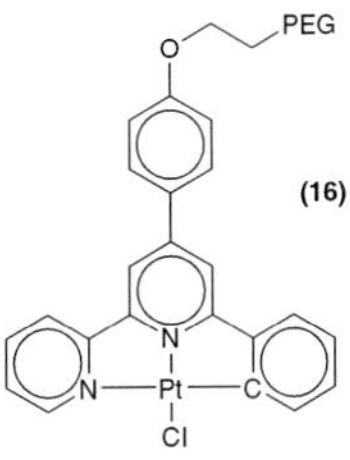

Fig. 16 An example of a Pt(II) complex with sensing properties for protein surfaces

containing Re(I) complexes (**14f** and **14g**, Fig. 14) [149] and the related Ir(III) complexes (**15d** and **15e**, Fig. 15) [150], demonstrate lifetime extension and enhanced fluorescence in the presence of oestrogen receptor α. In a related study, a cyclometallated Pt(II) complex (**16**, Fig. 16), bearing an extended poly(ethylene glycol) has demonstrated that the photoluminescence is enhanced in the presence of the hydrophobic regions of proteins such as BSA in aqueous solution [151].

6 Conclusions and Outlook

The versatility of organometallic complexes in optical sensing has been demonstrated by the variety of sensor compounds developed in recent years and the wide range of analytes that have been targeted. New research has built upon the properties of well-known chromophores, such as Re(I), Ir(III) or Ru(II) complexes, but also on less common species, such as Au–Tl/Ag networks or metal-containing self-assembled macrocycles, and has produced sophisticated architectures able of dual responses (e.g. optical and electrochemical) or containing various recognition sites appropriate for logic gates. Detailed systematic studies on the optical and structural properties of the complexes before and after inclusion of the analytes has brought a much better understanding of the mechanisms involved in the molecular recognition and reporting events. This fundamental research is essential to improve and predict the selectivity and sensitivity of the sensors, and further work is still needed if a rational design of sensors for specific target analytes is to be achieved. A further challenge will be the development of sensing devices with commercial potential. Whereas some examples of the incorporation of organometallic sensors into working optical devices have been reported, these are still scarce. However, given the increasing interest for reliable and cost-effective sensors for environmental and medical applications, this area is likely to receive much attention in coming years.

References

1. de Silva AP, Gunaratne HQN, Gunnlaugsson T, Huxley AJM, McCoy CP (1997) Chem Rev 97:1515–1566
2. Demas JN, DeGraff BA (2001) Coord Chem Rev 211:317–351
3. Keefe MH, Benkstein KD, Hupp JT (2000) Coord Chem Rev 205:201–228
4. Vos JG, Kelly JM (2006) Dalton Trans 4869–4883
5. Lowry MS, Bernhard S (2006) Chem Eur J 12:7970–7977
6. Wong KMC, Yam VWW (2007) Coord Chem Rev 251:2477–2488
7. Lazarides T, Miller TA, Jeffery JC, Ronson TK, Adams H, Ward MD (2005) Dalton Trans 528–536
8. Li MJ, Ko CC, Duan GP, Zhu NY, Yam VWW (2007) Organometallics 26:6091–6098
9. Lewis JD, Moore JN (2003) Chem Commun 2858–2859
10. Lewis JD, Moore JN (2004) Dalton Trans 1376–1385
11. Bakir M, Brown O, Johnson T (2004) J Mol Struct 691:265–272
12. Cattaneo M, Fagalde F, Katz NE (2006) Inorg Chem 45:6884–6891
13. Lazarides T, Easun TL, Veyne-Marti C, Alsindi WZ, George MW, Deppermann N, Hunter CA, Adams H, Ward MD (2007) J Am Chem Soc 129:4014–4027
14. Bergamini G, Saudan C, Ceroni P, Maestri M, Balzani V, Gorka M, Lee SK, van Heyst J, Vogtle F (2004) J Am Chem Soc 126:16466–16471
15. Li MJ, Chu BWK, Yam VWW (2006) Chem Eur J 12:3528–3537
16. Yam VWW, Ko CC, Chu BWK, Zhu NY (2003) Dalton Trans 3914–3921
17. Lai SW, Chan QKW, Zhu N, Che CM (2007) Inorg Chem 46:11003–11016
18. Konishi K, Yamaguchi H, Harada A (2006) Chem Lett 35:720–721
19. Zhao Q, Liu S, Shi M, Li F, Jing H, Yi T, Huang C (2007) Organometallics 26:5922–5930

20. Ho ML, Cheng YM, Wu LC, Chou PT, Lee GH, Hsu FC, Chi Y (2007) Polyhedron 26:4886–4892
21. Miyaji H, Collinson SR, Prokes I, Tucker JHR (2003) Chem Commun 64–65
22. Basurto S, Riant O, Moreno D, Rojo J, Torroba T (2007) J Org Chem 72:4673–4688
23. Pallavicini P, Dacarro G, Mangano C, Patroni S, Taglietti A, Zanoni R (2006) Eur J Inorg Chem 4649–4657
24. Lopez JL, Tarraga A, Espinosa A, Velasco MD, Molina P, Lloveras V, Vidal-Gancedo J, Rovira C, Veciana J, Evans DJ, Wurst K (2004) Chem Eur J 10:1815–1826
25. Tarraga A, Molina P, Lopez JL, Velasco MD (2004) Dalton Trans 1159–1165
26. Caballero A, Tarraga A, Velasco MD, Espinosa A, Molina P (2005) Org Lett 7:3171–3174
27. Oton F, Espinosa A, Tarraga A, Molina P (2007) Organometallics 26:6234–6242
28. Caballero A, Tarraga A, Velasco MD, Molina P (2006) Dalton Trans 1390–1398
29. Caballero A, Espinosa A, Tarraga A, Molina P (2008) J Org Chem 73:5489–5497
30. Caballero A, Espinosa A, Tarraga A, Molina P (2007) J Org Chem 72:6924–6937
31. Caballero A, Garcia R, Espinosa A, Tarraga A, Molina P (2007) J Org Chem 72:1161–1173
32. Delavaux-Nicot B, Maynadie J, Lavabre D, Fery-Forgues S (2007) J Organomet Chem 692:3351–3362
33. Delavaux-Nicot B, Maynadie J, Lavabre D, Fery-Forgues S (2006) Inorg Chem 45:5691–5702
34. Maynadie J, Delavaux-Nicot BM, Fery-Forgues S, Lavabre D, Mathieu R (2002) Inorg Chem 41:5002–5004
35. Maynadie J, Delavaux-Nicot B, Lavabre D, Fery-Forgues S (2006) J Organomet Chem 691:1101–1109
36. Zapata F, Caballero A, Espinosa A, Tarraga A, Molina P (2007) Org Lett 9:2385–2388
37. Caballero A, Tormos R, Espinosa A, Velasco MD, Tarraga A, Miranda MA, Molina P (2004) Org Lett 6:4599–4602
38. Yang QZ, Tong QX, Wu LZ, Wu ZX, Zhang LP, Tung CH (2004) Eur J Inorg Chem 1948–1954
39. Wong KMC, Tang WS, Lu XX, Zhu NY, Yam VWW (2005) Inorg Chem 44:1492–1498
40. Tang WS, Lu XX, Wong KMC, Yam VWW (2005) J Mater Chem 15:2714–2720
41. Han X, Wu LZ, Si G, Pan J, Yang QZ, Zhang LP, Tung CH (2007) Chem Eur J 13:1231–1239
42. Lanoë PH, Fillaut JL, Toupet L, Williams JAG, Le Bozec H, Guerchais V (2008) Chem Commun 4333–4335
43. Lo HS, Yip SK, Wong KMC, Zhu N, Yam VWW (2006) Organometallics 25:3537–3540
44. Resendiz MJE, Noveron JC, Disteldorf H, Fischer S, Stang PJ (2004) Org Lett 6:651–653
45. Yam VWW, Cheng ECC (2001) Gold Bull 34:20–23
46. Yam VWW, Cheung KL, Cheng ECC, Zhu NY, Cheung KK (2003) Dalton Trans 1830–1835
47. Yam VWW, Yip SK, Yuan LH, Cheung KL, Zhu NY, Cheung KK (2003) Organometallics 22:2630–2637
48. Li CK, Lu XX, Wong KMC, Chan CL, Zhu NY, Yam VWW (2004) Inorg Chem 43:7421–7430
49. de la Riva H, Nieuwhuyzen M, Fierro CM, Raithby PR, Male L, Lagunas MC (2006) Inorg Chem 45:1418–1420
50. Lagunas MC, Fierro CM, Pintado-Alba A, de la Riva H, Betanzos-Lara S (2007) Gold Bull 40:135–141
51. Tang HS, Zhu NY, Yam VWW (2007) Organometallics 26:22–25
52. Sun SS, Lees AJ (2002) Coord Chem Rev 230:171–192
53. Slone RV, Yoon DI, Calhoun RM, Hupp JT (1995) J Am Chem Soc 117:11813–11814
54. Sun SS, Anspach JA, Lees AJ, Zavalij PY (2002) Organometallics 21:685–693
55. Tzeng BC, Chen YF, Wu CC, Hu CC, Chang YT, Chen CK (2007) New J Chem 31:202–209
56. Beer PD, Dent SW, Hobbs GS, Wear TJ (1997) Chem Commun 99–100

57. Pelleteret D, Fletcher NC (2008) Eur J Inorg Chem 3597–3065
58. Pelleteret D, Fletcher NC, Doherty AP (2007) Inorg Chem 46:4386–4388
59. Sun SS, Lees AJ, Zavalij PY (2003) Inorg Chem 42:3445–3453
60. Sun SS, Lees AJ (2000) Chem Commun 1687–1688
61. Lin TP, Chen CY, Wen YS, Sun SS (2007) Inorg Chem 46:9201–9212
62. Uppadine LH, Redman JE, Dent SW, Drew MGB, Beer PD (2001) Inorg Chem 40:2860–2869
63. Beer PD, Timoshenko V, Maestri M, Passaniti P, Balzani V, Balzani B (1999) Chem Commun 1755
64. Curiel D, Beer PD, Paul RL, Cowley A, Sambrook MR, Szemes F (2004) Chem Commun 1162–1163
65. Curiel D, Beer PD (2005) Chem Commun 1909–1911
66. Lo KKW, Lau JSY, Fong VWY (2004) Organometallics 23:1098–1106
67. Fan Y, Zhu YM, Dai FR, Zhang LY, Chen ZN (2007) Dalton Trans 3885–3892
68. Fillaut JL, Andries J, Perruchon J, Desvergne JP, Toupet L, Fadel L, Zouchoune B, Saillard JY (2007) Inorg Chem 46:5922–5932
69. Lo KKW, Lau JSY, Lo DKK, Lo LTL (2006) Eur J Inorg Chem 4054–4062
70. Beer PD, Bayly SR (2005) Top Curr Chem 255:125–162
71. Beer PD, Hayes EJ (2003) Coord Chem Rev 240:167–189
72. Beer PD, Graydon AR, Sutton LR (1996) Polyhedron 15:2457–2461
73. Beer PD, Szemes F, Balzani V, Sala CM, Drew MGB, Dent SW, Maestri M (1997) J Am Chem Soc 119:11864–11875
74. Oton F, Tarraga A, Molina P (2006) Org Lett 8:2107–2110
75. Oton F, Tarraga A, Espinosa A, Velasco MD, Molina P (2006) J Org Chem 71:4590–4598
76. Huynh L, Wang ZU, Yang J, Stoeva V, Lough A, Manners I, Winnik MA (2005) Chem Mat 17:4765–4773
77. Albrecht M, van Koten G (2001) Angew Chem Int Ed 40:3750–3781
78. Cabanillas-Galan P, Farmer L, Hagan T, Nieuwenhuyzen M, James SL, Lagunas MC (2008) Inorg Chem 47:9035–9041
79. Kato M (2007) Bull Chem Soc Jpn 80:287–294
80. Isci H, Mason WR (1974) Inorg Chem 13:1175–1180
81. Exstrom CL, Sowa JR, Daws CA, Janzen D, Mann KR, Moore GA, Stewart FF (1995) Chem Mater 7:15–17
82. Bailey RC, Hupp JT (2002) J Am Chem Soc 124:6767–6774
83. Grate JW, Moore LK, Janzen DE, Veltkamp DJ, Kaganove S, Drew SM, Mann KR (2002) Chem Mat 14:1058–1066
84. Drew SM, Mann JE, Marquardt BJ, Mann KR (2004) Sens Actuator B Chem 97:307–312
85. Buss CE, Mann KR (2002) J Am Chem Soc 124:1031–1039
86. Dylla AG, Janzen DE, Pomije MK, Mann KR (2007) Organometallics 26:6243–6247
87. Sun Y, Ye K, Zhang H, Zhang J, Zhao L, Li B, Yang G, Yang B, Wang Y, Lai SW, Che CM (2006) Angew Chem Int Ed 45:5610–5613
88. Kato M, Kishi S, Wakamatsu Y, Sugi Y, Osamura Y, Koshiyama T, Hasegawa M (2005) Chem Lett 34:1368–1369
89. Fornies J, Fuertes S, Lopez JA, Martin A, Sicilia V (2008) Inorg Chem 47:7166–7176
90. Lu W, Chan MCW, Zhu NY, Che CM, He Z, Wong KY (2003) Chem Eur J 9:6155–6166
91. Kui SCF, Chui SSY, Che CM, Zhu NY (2006) J Am Chem Soc 128:8297–8309
92. Vickery JC, Olmstead MM, Fung EY, Balch AL (1997) Angew Chem Int Ed Eng 36:1179–1181
93. Fernandez EJ, Lopez-De-Luzuriaga JM, Monge M, Montiel M, Olmos ME, Perez J, Laguna A, Mendizabal F, Mohamed AA, Fackler JP (2004) Inorg Chem 43:3573–3581
94. Fernandez EJ, Lopez-de-Luzuriaga JM, Monge M, Olmos ME, Perez J, Laguna A, Mohamed AA, Fackler JP (2003) J Am Chem Soc 125:2022–2023

95. Fernandez EJ, Laguna A, Lopez-de-Luzuriaga JM, Montiel M, Olmos ME, Perez J (2006) Organometallics 25:1689–1695
96. Fernandez EJ, Lopez-De-Luzuriaga JM, Monge M, Olmos ME, Puelles RC, Laguna A, Mohamed AA, Fackler JP (2008) Inorg Chem 47:8069–8076
97. Luquin A, Bariain C, Vergara E, Cerrada E, Garrido J, Matias IR, Laguna M (2005) Appl Organomet Chem 19:1232–1238
98. Bariain C, Matias IR, Fdez-Valdivielso C, Elosua C, Luquin A, Garrido J, Laguna M (2005) Sens Actuator B Chem 108:535–541
99. Elostua C, Bariain U, Matias JR, Arregui FJ, Luquin A, Laguna M (2006) Sens Actuator B Chem 115:444–449
100. Elosua C, Matias IR, Bariain C, Arregui FJ (2006) Sensors 6:1440–1465
101. Terrones SC, Aguado CE, Bariain C, Carretero AS, Maestro IRM, Gutierrez AF, Luquin A, Garrido J, Laguna M (2006) Opt Eng 45
102. Luquin A, Elosua C, Vergara E, Estella J, Cerrada E, Bariain C, Matias IR, Garrido J, Laguna M (2007) Gold Bull 40:225–233
103. Leznoff DB, Lefebvre J (2005) Gold Bull 38:47–54
104. Katz MJ, Ramnial T, Yu HZ, Leznoff DB (2008) J Am Chem Soc 130:10662–10673
105. Liu ZW, Bian ZQ, Bian J, Li ZD, Nie DB, Huang CH (2008) Inorg Chem 47:8025–8030
106. Slone RV, Hupp JT (1997) Inorg Chem 36:5422–5423
107. Thanasekaran P, Liao RT, Manimaran B, Liu YH, Chou PT, Rajagopal S, Lu KL (2006) J Phys Chem A 110:10683–10689
108. Lee SJ, Lin WB (2002) J Am Chem Soc 124:4554–4555
109. Müller C, Whiteford JA, Stang PJ (1998) J Am Chem Soc 120:9827–9837
110. Yam VWW, Kai ASF (1998) Chem Commun 109–110
111. Mizuno T, Fukumatsu T, Takeuchi M, Shinkai S (2000) J Chem Soc Perkin Trans 1:407–413
112. Cary DR, Zaitseva NP, Gray K, O'Day KE, Darrow CB, Lane SM, Peyser TA, Satcher JH, Van Antwerp WP, Nelson AJ, Reynolds JG (2002) Inorg Chem 41:1662–1669
113. Chow CF, Chiu BKW, Lam MHW, Wong WY (2003) J Am Chem Soc 125:7802–7803
114. Chow CF, Lam MHW, Sui HY, Wong WY (2005) Dalton Trans 475–484
115. Chen HL, Zhao Q, Wu YB, Li FY, Yang H, Yi T, Huang CH (2007) Inorg Chem 46:11075–11081
116. Erkkila KE, Odom DT, Barton JK (1999) Chem Rev 99:2777–2795
117. Metcalfe C, Thomas JA (2003) Chem Soc Rev 32:215–224
118. Pierard F, Kirsch-De Mesmaeker A (2006) Inorg Chem Commun 9:111–126
119. Thornton NB, Schanze KS (1996) New J Chem 20:791–800
120. Thornton NB, Schanze KS (1993) Inorg Chem 32:4994–4995
121. Stoeffler HD, Thornton NB, Temkin SL, Schanze KS (1995) J Am Chem Soc 117:7119–7128
122. Yam VW-W, Lo KK-W, Cheung K-K, Kong RY-C (1997) J Chem Soc Dalton Trans 2067–2072
123. Ruiz GT, Juliarena MP, Lezna RO, Wolcan E, Feliz MR, Ferraudi G (2007) Dalton Trans 2020–2029
124. Metcalfe C, Webb M, Thomas JA (2002) Chem Commun 2026–2027
125. Foxon SP, Phillips T, Gill MR, Towrie M, Parker AW, Webb M, Thomas JA (2007) Angew Chem Int Ed 46:3686–3688
126. Lo KKW, Hui WK, Chung CK, Tsang KHK, Ng DCM, Zhu NY, Cheung KK (2005) Coord Chem Rev 249:1434–1450
127. Lo KKW, Ng DCM, Chung CK (2001) Organometallics 20:4999–5001
128. Lo KKW, Chung CK, Zhu NY (2003) Chem Eur J 9:475–483
129. Lo KKW, Chung CK, Lee TKM, Lui LH, Tsang KHK, Zhu NY (2003) Inorg Chem 42:6886–6897
130. Lo KKW, Li CK, Lau KW, Zhu NY (2003) Dalton Trans 4682–4689

131. Wong KMC, Tang WS, Chu BWK, Zhu NY, Yam VWW (2004) Organometallics 23:3459–3465
132. Dattelbaum JD, Abugo OO, Lakowicz JR (2000) Bioconjug Chem 11:533–536
133. Guo XQ, Castellano FN, Li L, Szmacinski H, Lakowicz JR, Sipior J (1997) Anal Biochem 254:179–186
134. Hamzavi R, Happ T, Weitershaus K, Metzler-Nolte N (2004) J Organomet Chem 689:4745–4750
135. Lo KKW, Ng DCM, Hui WK, Cheung KK (2001) J Chem Soc Dalton Trans 2634–2640
136. Garino C, Ghiani S, Gobetto R, Nervi C, Salassa L, Ancarani V, Neyroz P, Franklin L, Ross JBA, Seibert E (2005) Inorg Chem 44:3875–3879
137. Dunn AR, Belliston-Bittner W, Winkler JR, Getzoff ED, Stuehr DJ, Gray HB (2005) J Am Chem Soc 127:5169–5173
138. Lo KKW, Tsang KHK, Sze KS, Chung CK, Lee TKM, Zhang KY, Hui WK, Li CK, Lau JSY, Ng DCM, Zhu NY (2007) Coord Chem Rev 251:2292–2310
139. Lo KKW, Tsang KHK, Sze KS (2006) Inorg Chem 45:1714–1722
140. Lo KKW, Hui WK (2005) Inorg Chem 44:1992–2002
141. Lo KKW, Tsang KHK (2004) Organometallics 23:3062–3070
142. Lo KKW, Li CK, Lau JSY (2005) Organometallics 24:4594–4601
143. Lo KKW, Lau JSY (2007) Inorg Chem 46:700–709
144. Lo KKW, Hui WK, Ng DCM (2002) J Am Chem Soc 124:9344–9345
145. Lo KKW, Chung CK, Zhu NY (2006) Chem Eur J 12:1500–1512
146. Lo KKW, Louie MW, Sze KS, Lau JSY (2008) Inorg Chem 47:602–611
147. Lo KKW, Tsang KHK, Hui WK, Zhu N (2005) Inorg Chem 44:6100–6110
148. Lo KKW, Sze KS, Tsang KHK, Zhu NY (2007) Organometallics 26:3440–3447
149. Lo KKW, Tsang KHK, Zhu NY (2006) Organometallics 25:3220–3227
150. Lo KKW, Zhang KY, Chung CK, Kwok KY (2007) Chem Eur J 13:7110–7120
151. Che CM, Zhang JL, Lin LR (2002) Chem Commun 2556–2557

Metal Complexes Featuring Photochromic Ligands

Véronique Guerchais and Hubert Le Bozec

Abstract Organic photochromic molecules are important for the design of photoresponsive functional materials such as switches and memories. Over the past 10 years, research efforts have been directed towards the incorporation of photoresponsive molecules into metal systems, in order either to modulate the photochromic properties or to photoregulate the redox, optical, and magnetic properties of the organometallic moieties. This chapter focuses on work reported within the last few years in the area of organometallic and coordination complexes containing photochromic ligands. The first part is related to photochromic azo-containing metal complexes. The second part deals with metal complexes incorporating 1,2-dithienylethene (DTE) derivatives. The last three parts are devoted to metal complexes featuring other photochromes such as spiropyran, spirooxazine, benzopyran, and dimethyldihydropyrene derivatives.

Keywords Photochromes, Transition metal complexes

Contents

1 Introduction 172
2 Photochromic Azo-Containing Metal Complexes 173
   2.1 Introduction 173
   2.2 Azoferrocene and Ferrocene–Azobenzene Derivatives 173
   2.3 Metal Complexes with Azobenzene-Conjugated Bipyridine Ligands 178
   2.4 Metal Complexes with Azobenzene-Conjugated Terpyridine Ligands 182
   2.5 Metalladithiolenes with Azobenzene Groups 184
   2.6 Azobenzene-Containing Metal Alkynyl Complexes 186
3 Metal Complexes Incorporating 1,2-Dithienylethene 188
   3.1 Introduction 188
   3.2 Organo-Boron DTE-Based Dithienylcyclopentene 189

V. Guerchais and H. Le Bozec (✉)
Laboratoire de Sciences Chimiques de Rennes, UMR 6226 Université de Rennes1-CNRS, Campus de Beaulieu, 35042 Rennes cedex, France
e-mail: hubert.le-bozec@univ-rennes1.fr

3.3	Complexes Incorporating DTE-Based Pyridine, Cyano, or Carboxylate Ligands	191
3.4	Photoregulation of Luminescence	198
3.5	Photoswitching of Second-Order NLO Activity	203
3.6	Photoswitching of a Magnetic Interaction	204
3.7	Photo- and Electrochromic Properties of Metal-Based DTE Derivatives	205
3.8	Other DTE-Based Metal Complexes	207
3.9	DTE-Based Ligands in Catalysis	208
3.10	Multi-DTE Metal Complexes	209
4	Photochromic Spiropyran and Spirooxazine-Containing Metal Complexes	210
4.1	Introduction	210
4.2	Ferrocenylspiropyran	211
4.3	Porphyrin Spiropyran Metal Complexes	212
4.4	Spiropyran- and Spirooxazine-Containing Polypyridine Metal Complexes	212
4.5	Diastereomeric Isomerism in [(η^6-Spirobenzopyran)Ru(C5Me5)]$^+$	216
4.6	Complexes of Spiropyrans with Metal Ions	216
5	Photochromic Metallocenyl Benzopyran Derivatives	217
6	Dimethyldihydropyrene Metal Complexes	218
7	Other Photochromic Metal Complexes	220
7.1	Terthiazole Derivatives	221
8	Conclusion	221
References		222

1 Introduction

Photochromism, which refers to the reversible color change of a compound with light irradiation, is attracting much attention for the construction of molecular devices. During the past decade there has been a growing interest in the synthesis, properties, and applications of organic photochromic materials [1–5].

Photochromic materials have been the focus of intensive investigations for several decades from both the fundamental and practical points of view for their potential applications to optically rewritable data storage, optical switching, and chemical sensing.... Useful properties that may be photoregulated include luminescence, refractive index, electronic conductance, magnetism, optical rotation, nonlinear optics, redox chemistry.... Photochromic transformations are generally based on unimolecular processes involving the interconversion of two isomers, such as *cis/trans* isomerization, ring opening/closing, or intramolecular proton transfer. So far, various types of organic photochromic compounds such as azobenzenes, diarylethenes, fulgides, spirobenzopyrans, and dimethyldihydropyrenes have been developed.

Metal complexes featuring photoresponsive ligands are an interesting alternative to pure organic photochromes. Combining a photochromic moiety with an organometallic or coordination compound will provide new properties deriving from the combination of redox, optical and magnetic properties of the metal complexes with the photochromic reaction.

The aim of the present chapter is to review some recent developments, made in the last 10 years in organometallic and coordination compounds that contain ligands functionalized by organic photochromic units such as azobenzene, dithienylethene (DTE), spiropyran (SP) and spirooxaxine (SO), benzopyran, and dimethyldihydropyrene, respectively. Other known types of photochromic metal complexes based on linkage isomerization of coordinated ligands such as dimethyl sulfoxide and pyridine will not be discussed in this chapter.

2 Photochromic Azo-Containing Metal Complexes

2.1 Introduction

Azobenzene derivatives constitute a family of dyes which are well known for their photochromic properties. The photochemistry of azobenzene derivatives has been extensively studied in solution as well in polymer matrices [6, 7]. These compounds have been widely investigated as promising systems for various applications such as photo-switching devices, optical data storage, holography, and nonlinear optics [8, 9]. Basically, they are characterized by a *trans* to *cis* isomerization of the N=N double bond upon UV light irradiation, and the reverse isomerization can take place by visible light irradiation or by heating, the *trans* form being generally more stable than the *cis* form. Azo-containing transition metal complexes can provide interesting versatile molecular materials because of the combination of the photoisomerization behavior of the azo group with the optical, redox, and magnetic properties of the metal complexes [10–12].

2.2 Azoferrocene and Ferrocene–Azobenzene Derivatives

The synthesis of azoferrocene (azoFc), one of the simplest organometallic analogs of azobenzene, was reported in 1961 by Nesmeyanov [13, 14]. Nishihara et al. examined much later the isomerization behavior of this interesting bimetallic complex [15]. The UV–visible spectrum of *trans*-azoFc in acetonitrile shows two strong absorption bands at 318 and 530 nm assigned to π–π* transition of the azo group and MLCT ($d_{Fe} \rightarrow \pi^*_{CpN=NCp}$) transition, respectively. Upon UV irradiation, the π–π* band decreased and a new band appeared at 368 nm, showing isosbestic points. This new band was assigned to the n–π* transition of the *cis* isomer. Remarkably, the photoisomerization proceeded upon irradiation in the MLCT band ($\lambda = 546$ nm). This is a rare example of photoisomerization using a much longer wavelength than that of the π–π* band. However, the back *cis*→*trans* reaction could not be observed by either heating or irradiation with visible light.

trans-azoFc → hv 365 or 546 nm → cis-azoFc

In order to study the photoisomerization using MLCT, Nishihara prepared a series of azobenzene derivatives substituted by a ferrocenyl group at the *meta* position of one phenyl ring [16, 17]. For example, 3-ferrocenylazobenzene **1** showed an interesting redox-conjugated reversible isomerization cycle using a single light.

1 R = H, R' = H
2 R = Cl, R' = H
3 R = H, R' = CO$_2$H

Compound **1** displays in acetonitrile a strong azo π–π* band at 318 nm and a weaker MLCT band at 444 nm. The *trans* to *cis* photoisomerization proceeded upon either UV light (320 nm) or green light (546 nm) irradiation, at wavelengths corresponding to the maximum of the π–π* band and to the edge of MLCT band, respectively. The *cis* molar ratio in the photostationary state (PSS) upon UV light and green light irradiation were estimated to be 61% and 35%, respectively. Upon chemical or electrochemical one electron oxidation to the resulting **1$^+$**, the MLCT band disappeared, and a new weak ligand-to-metal charge transfer (LMCT) band appeared at 730 nm. This ferrocenium complex showed only *trans* to *cis* photoisomerization with UV light irradiation, whereas the reverse *cis* to *trans* reverse isomerization proceeded by excitation of the n–π* band with green light. The different responses to green light between the Fe(II) and Fe(III) states were used to achieve a reversible photoisomerization with a single visible light source by combination with the reversible redox Fe^{III}/Fe^{II} reaction (Scheme 1).

In order to improve the efficiency of photoisomerization with green light, different ferrocenyl azobenzene derivatives were also prepared either by adding substituents on the benzene ring or by changing the position of the ferrocenyl moiety (*ortho*, *meta* or *para*) on the benzene ring [17]. Upon introducing an electron-withdrawing substituent, such as a chloro group in *para* position to the ferrocenyl group (**2**), the *cis* molar ratio at 546 nm light irradiation was increased from 35% (R = H) to 47% (R = Cl). More recently, Nishihara et al. succeeded

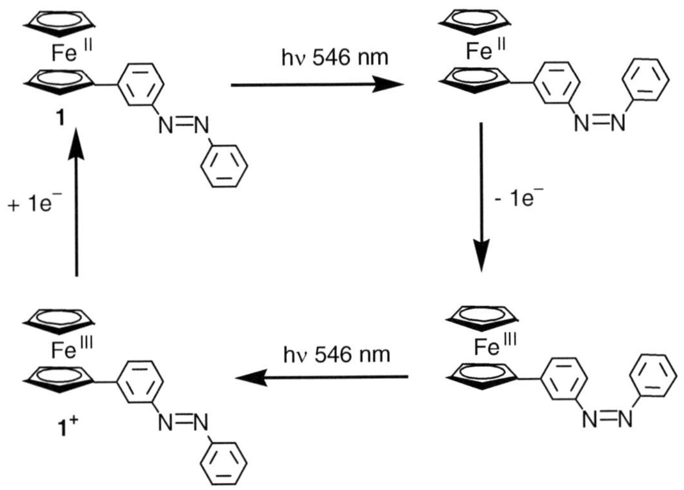

Metal Complexes Featuring Photochromic Ligands

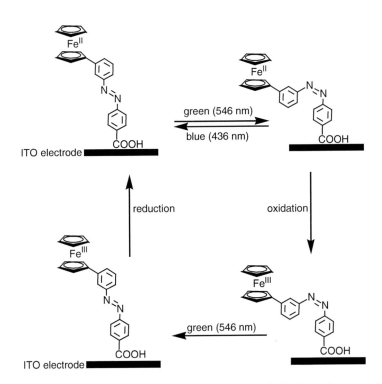

Scheme 1 Reversible photoisomerization of **1** with single light and redox reaction

Scheme 2 Schematic representation of the couple redox-single light isomerization of **3**/ITO

in constructing a single light controllable azobenzene monolayer system by electrochemical oxidation [18]. They synthesized the 3-ferrocenyl-4'-carboxylazobenzene compound **3** containing a carboxylic acid end group, and then prepared a monolayer of **3** on a transparent indium tin oxide (ITO) electrode. The photoswitching behavior of this system was nicely evidenced and, like the parent 3-ferrocenylazobenzene **1** in solution, the single-light photoisomerization cycle of **3**/ITO was achieved by the combination of green light and electrochemical reactions (Scheme 2).

Aida and coworkers have developed novel molecular machines containing a photochromic azobenzene unit and a ferrocenyl moiety. They have designed an interesting "light-driven chiral molecular scissors" **4**, consisting of a tetrasubstituted ferrocene as the pivot part which is able to generate an angular motion, two phenyl groups as the blade moieties, and two phenylene groups as the handle parts strapped by an photoisomerizable azobenzene unit through ethylene linkage [19].

UV light irradiation at 350 nm of a THF solution of the *trans* isomer resulted in a typical *trans* to *cis* isomerization of the azobenzene moiety (89% conversion), whereas irradiation with a visible light gave back to a *trans/cis* isomer ratio of 46:54. Absorption, circular dichroism (CD) and ^{1}H NMR spectra of an enantiomer of **4** agreed well with the prediction by DFT calculations that the blade parts are opened when the azobenzene unit adopts a *cis* configuration while they are closed in the case of the *trans* isomer. Interestingly, the oxidation state of the ferrocene pivot was found to affect the PSSs of **4** and thus to allow a reversible open–closed motion by use of redox and UV light [20] (Scheme 3).

This concept was then nicely applied by Aida to develop light-powered molecular pliers **5** that can bind and deform guest molecules [21, 22]. They introduced an appropriate guest-binding site – a zinc porphyrin complex – at each cyclopentadienyl ring of the ferrocene. The zinc porphyrin units were found to bind bidentate guests such as 4,4'-biisoquinoline, forming a stable host–guest complex. Upon exposition to UV and visible light, *trans/cis* photoisomerization of azobenzene induced mechanical twisting of the guest molecule, which was detected by changes in the CD spectra (Scheme 4).

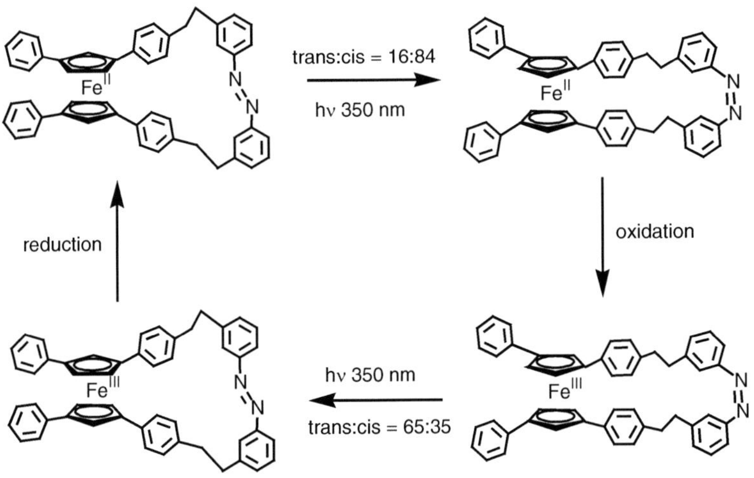

Scheme 3 Open/closed motion of **4** induced by redox and UV light

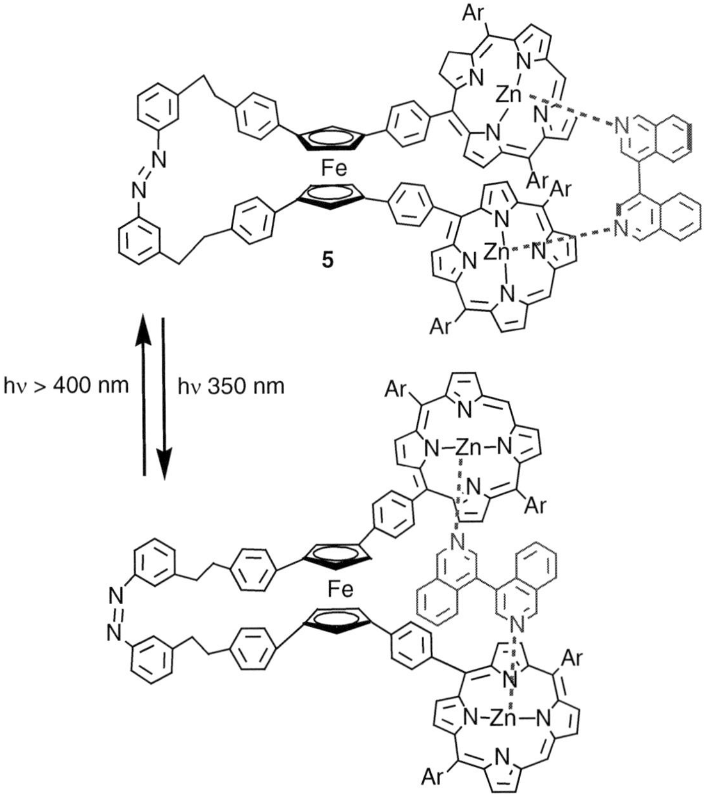

Scheme 4 Photoisomerization of guest–host molecular pliers upon UV and visible irradiation

2.3 Metal Complexes with Azobenzene-Conjugated Bipyridine Ligands

2.3.1 Tris(bipyridine) Metal Complexes

The first example of reversible *trans–cis* isomerization of the azo group achieved through a combination of photoirradiation and a redox cycle between Co(II) and Co (III) was reported by Nishihara [23]. The tris(*p*-azobipy)Co(II) complex used in this system was prepared from 4-tolylazophenyl-2,2'-bipyridine (*p*-azobipy) and Co^{II}(NO$_3$)$_2$ and characterized by a reversible Co(III)/Co(II) wave at low potential (-0.15 V vs Fc$^+$/Fc in dichloromethane). The corresponding Co(III) complex was readily obtained upon oxidation of the Co(II) complex with silver triflate. Both Co (II) and Co(III) complexes displayed in UV–visible spectroscopy a strong $\pi-\pi^*$ band due to the azo group at $\lambda = 360$ nm. Upon irradiation, the two complexes showed different behavior: UV light irradiation at 366 nm of a CH$_2$Cl$_2$ solution of [Co^{II}(*p*-azobipy)$_3$](BF$_4$)$_2$ resulted in a decrease of the $\pi-\pi^*$ band and an increase of the $n-\pi^*$ band at ca. 450 nm, in agreement with a typical *trans*-to-*cis* isomerization of the azobenzene moiety (40% conversion at the PSS), whereas irradiation with 438 nm light gave back the *trans* isomer. By contrast, almost no decrease in absorbance of $\pi-\pi^*$ band of [Co^{III}(*p*-azobipy)$_3$](BF$_4$)$_3$ could be observed under UV light irradiation, suggesting a much more effective *cis*-to-*trans* back reaction in the case of the oxidized Co(III) complex. The same behavior was observed in the case of [Co^{II}(*m*-azobipy)$_3$](BF$_4$)$_2$ and [Co^{III}(*m*-azobipy)$_3$](BF$_4$)$_3$ which showed in the PSS 57% and 9% *trans* to *cis* photoconversion, respectively [24].

The difference in *cis* form concentrations between Co(II) and Co(III) was applied for the reversible photoisomerization of the azobenzene moieties upon excitation with a single UV light source and addition of stoichiometric chemical oxidizing and reducing agents, respectively (Scheme 5) [23, 24].

Tris(styrylbipyridine)zinc(II) complexes functionalized by dialkylamino-azobenzene groups have recently been designed by Le Bozec et al. to prepare photoisomerizable star-shaped nonlinear optical polymers **7** [25].

Recent studies have highlighted the potential of bipyridyl metal complexes in the field of nonlinear optics, and high molecular hyperpolarizability β values have been reported by our group for octupolar (D_3) tris(4,4'-disubstituted-2,2'-bipyridine) metal

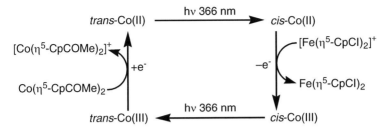

Scheme 5 Reversible *trans-cis* isomerization of an azo group through a combination of UV light and Co(III)/Co(II) redox change

complexes [26]. Because the traditional electric-field poling-which is the relevant technique for the molecular orientation of dipolar chromophores-is not applicable for octupolar NLO-phores due to the absence of permanent dipole moment, the so-called "all optical poling" technique has been used to induce noncentrosymmetric ordering of multipolar molecules in polymer films. Basically, this method requires the use of polymer matrices containing NLO chromophores featuring photoisomerizable moieties such as an azobenzene group which can undergo reversible *trans/cis/trans* photoisomerization cycles. In chloroform solution, ligand **6** and complex **7** showed the usual photoisomerization behavior of push-pull azobenzene derivatives [25, 27]. The corresponding photoisomerizable star-shaped polymer was subsequently prepared by atom transfer radical polymerization (ATRP) of methylmethacrylate (MMA). The resulting polymer film also exhibited a photoisomerization behavior typical of azo dyes. Finally, the macroscopic optical molecular orientation of this grafted NLO-polymer film was promoted for the first time using a combined one- and two-photon excitation at 1,064 nm and 532 nm, respectively [28].

2.3.2 Bis(bipyridine) Metal Complexes

The key feature to stabilize pseudotetrahedral Cu(I) complexes of ligands such as 2,2-bipy is the incorporation of substituents α to the imine nitrogen.

In contrast, in the absence of *ortho* substituents, Cu(I) complexes are readily oxidized to the more stable square planar Cu(II) complexes. Nishihara et al. used this coordination motion to promote the reversible *trans-cis* photoisomerization of an azobenzene-attached 6,6′dimethyl-2,2′-bipyridine (dmbipy-azo) with a single light source, by controlling the binding/release reaction of the ligand to copper [29]. Upon irradiation at 365 nm, the free *trans*-dmbipy-azo ligand was almost quantitatively converted to the *cis* isomer. On the other hand, the *cis* molar ratio at PSS were only 18% and 14% for the corresponding Cu(I) and Cu(II) complexes, respectively. A cyclic voltammetry study of [(dmbipy-azo)$_2$Cu]$^+$ in the presence of 2 equivalents of 2,2′-bipy also revealed a facile ligand exchange reaction between the two bipyridine ligands, the driving force for this exchange being the difference in coordination geometry between Cu(I) and Cu(II) (Scheme 6).

By combining the dynamic motion of the metal complex with that of the azobenzene ligand, an efficient reversible *trans–cis* isomerization of [(dmbipy-azo)$_2$Cu]$^+$ upon UV irradiation was observed with an increase of the *cis/trans* ratio to 70%. The experiment was performed chemically in the presence of 2 equivalents of bipy with oxidizing and reducing agents (Scheme 7).

This concept of reversible ligand exchange reaction in pseudotetrahedral Cu(I) complexes was also exploited by Nishihara and coworkers to convert

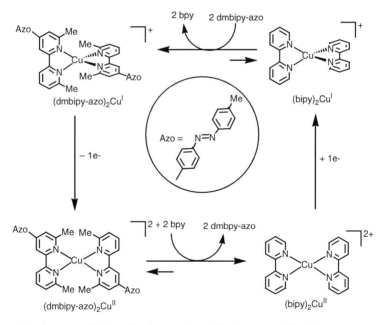

Scheme 6 Redox-controlled coordination reaction of dmbpy-azo and bpy ligands

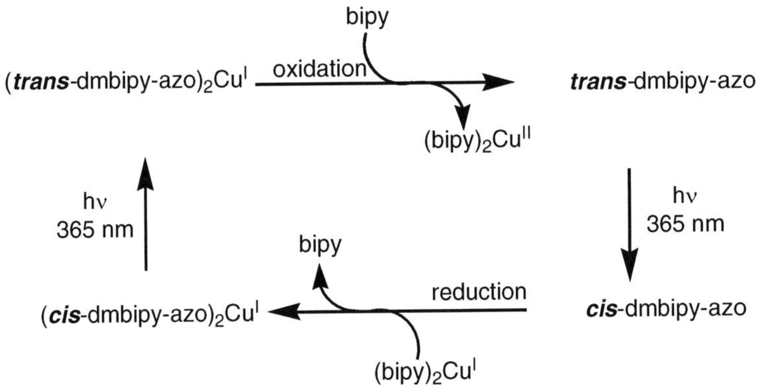

Scheme 7 Coordination synchronized *trans-cis* photoisomerization of dmbipy-azo driven by Cu(II)/Cu(I) redox change

photon energy into electronic potential energy [30]. They developed a new 2,2′-bipyridine ligand *trans₂-o*-AB containing two azobenzene moieties at the 6,6′ position and prepared the corresponding [Cu(*trans₂-o*-AB)₂]⁺ and [Cu(*cis₂-o*-AB)₂]⁺ complexes, the latter being formed upon photoisomerization with UV light of the *trans₂-o*-AB ligand.

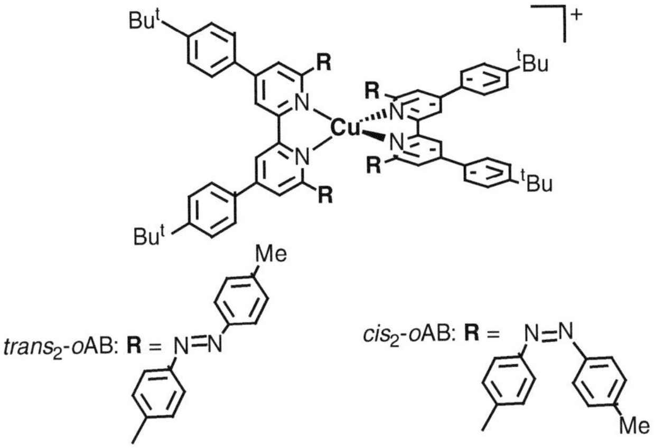

Due to interligand π-stacking effect, the coordination structure of [Cu(*trans₂-o*-AB)₂]⁺ was found to be much more stable than that of [Cu(*cis₂-o*-AB)₂]⁺. The destabilization of the *cis* isomer was used to drive a ligand exchange reaction between *cis₂-o*-AB and 2,2′-bipy, and concomitantly to shift the Cu^II/Cu^I redox potential by ca. 0.6 V. The combination of the *trans/cis* photoisomerization of azobenzene with the ligand exchange reaction was nicely applied to construct an

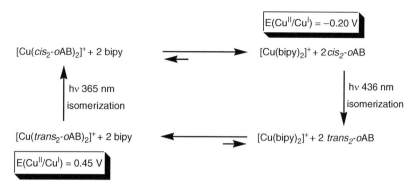

Scheme 8 Photoelectric conversion induced by a ligand exchange reaction

artificial molecular machine where UV/visible light information can be transformed into an electrode change (Scheme 8).

2.4 Metal Complexes with Azobenzene-Conjugated Terpyridine Ligands

The photoisomerization properties of mononuclear and dinuclear transition metal complexes containing azobenzene-conjugated terpyridine ligands (*trpy–azo* and *trpy–azo–trpy*) have been investigated with several d^6 (Fe^{II}, Ru^{II}, Co^{III}, Rh^{III}) and d^7 (Co^{II}) metal ions [31–34]. The *trans/cis* photoisomerization behavior was found to be strongly dependent on the nature of the metal, as well on the nature of the counter anion and the solvent. Upon UV irradiation, the monometallic ruthenium complex [(trpy)Ru(*trpy–azo*)]$^{2+}$ underwent *trans* to *cis* photoisomerization to reach a PSS with only 20% of the *cis* form, whereas the bimetallic [(trpy)Ru(*trpy–azo–trpy*)Ru(trpy)]$^{4+}$ did not photoisomerize at all [31, 33]. In contrast, the corresponding mono- and bimetallic Rh(III) complexes were found to isomerize almost totally [32, 33]. The reverse *cis* to *trans* isomerization upon visible light irradiation was

M = Ru^II, Co^II : n = 4
Co^III, Rh^III : n = 6

X^- = BF_4^-, PF_6^-, BPh_4^-

not observed for both rhodium compounds but only upon heating in the dark. The absence of *trans* to *cis* photoisomerization in the case of Ru complexes was attributed to energy transfer from the azo excited state to the low-lying MLCT state. As for Ru complexes, the monometallic Fe complex showed an MLCT band at low energy [λ_{MLCT} (CH$_3$CN) = 580 nm], and the *cis* to *trans* photoisomerization behavior was depressed by an energy transfer pathway [34].

Interestingly, the rate of the *trans* to *cis* photoisomerization of Rh complexes was found to depend on the nature of the counter anion: for example, the use of BPh$_4^-$ instead of PF$_6^-$ or BF$_4^-$ resulted in a much faster photoisomerization [32, 33]. This difference was attributed to the larger size of BPh$_4^-$ which prevents strong ion pairing with the complex cation, and consequently reduces the apparent rotor volume, resulting in a faster photoisomerization. The photochemical properties of the mono- and bimetallic d^7 Co(II) and d^6 Co(III) complexes were also investigated [34]. The Co(III) complexes were easily obtained upon oxidation of the Co(II) compounds with silver salts. The Co(II) complexes exhibit reversible *trans/cis/trans* photoisomerization of the azo group in propylene carbonate (PC) upon irradiation at 366 nm and 435 nm, respectively. The *cis* to *trans* isomerization is also effective in PC by heating at 90°C. The *trans* to *cis* photoisomerization of the Co(III) complexes also occurred at 366 nm in dichloroethane, although less efficiently than that of Co(II), but the back photoisomerization could not be observed by either visible irradiation or heat. Thus these studies demonstrated the profound influence of the metal center, as well as of the oxidation state on the photochromic behavior of the azobenzene unit.

The photoluminescence switching upon *trans–cis* photoisomerization of the azobenzene moiety of square planar platinum(II) complexes **8** and **9** were investigated by Nishihara and coworkers [35].

8 : L = py, n = 2
9 : L = Cl, n = 1

hν (UV)
hν (visible) or heat (100°C)

emission OFF (λ_{exc} = 450 nm)

emission ON
(λ_{exc} = 450 nm, λ_{em} = 600 nm)

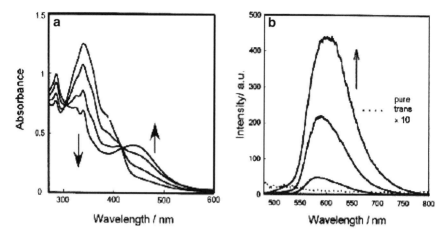

Fig. 1a UV–visible absorption change of **9** in propylene carbonate upon irradiation at 366 nm. **b** Emission spectral change of **9** in EtOH–MeOH–DMF at 77 K upon irradiation at 366 nm ($\lambda_{exc} = 450$ nm) (Reprinted with permission from [35])

Their absorption spectra show at ca. 350 nm characteristic strong azo $\pi-\pi^*$ bands overlapped with weaker MLCT bands. Upon photoirradiation at 366 nm, efficient *trans* to *cis* photoisomerization occurred as evidenced by the decrease of the $\pi-\pi^*$ band and the apparition of the azo $n-\pi^*$ band at around 450 nm (Fig. 1).

The reverse *cis* to *trans* isomerization was observed either by irradiation with visible light or by heat. No emission in solution at room temperature could be observed for both isomers. In organic glasses at 77 K a dramatic emission spectral enhancement was observed by the *cis–trans* conformation change: whereas the *trans* forms are nonluminescent at excitation wavelength of 450 nm, the *cis* isomers exhibit a red luminescence centered at 600 nm. The long emission lifetime of 40 μs is typical of an ^{3}MLCT excited state, probably with mixing of some $^3\pi-\pi^*$ excited state. This OFF/ON switching was interpreted by the nonplanar geometry of the *cis* form and the reduced π-conjugation effect.

2.5 Metalladithiolenes with Azobenzene Groups

Metal dithiolene complexes represent an important class of organometallic molecules which have been widely studied for their interesting properties such as redox activity, conductivity, and magnetism. In order to design new photo-responsive multifunctional compounds, a series of complexes combining metalladithiolenes with azobenzene groups has been described by Nishihara et al. [36–38]. New square planar azobenzodithiolene Ni, Pd, and Pt complexes were prepared from the benzodithiolethione precursor [36, 37].

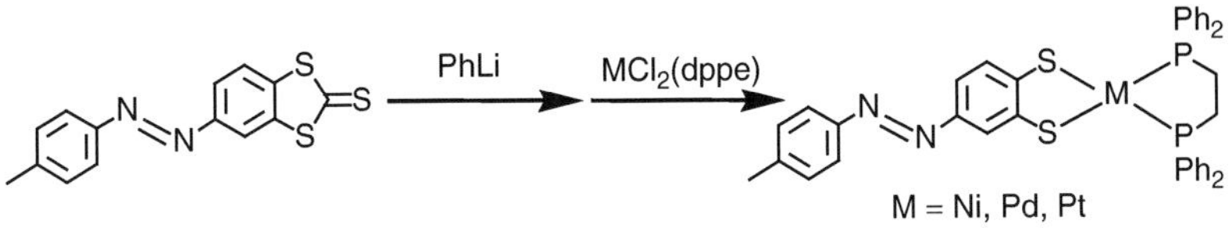

Upon photoirradiation in the π–π* transition band at 405 nm, a decrease of this band was observed, indicating the occurrence of the *trans* to *cis* photoisomerization. A conversion of ca. 40% in PSS was found, according to ^{1}H NMR spectroscopic measurements. The back *cis* to *trans* photoisomerization also occurred upon photoirradiation, but unusually by using UV light excitation, with an energy higher than the π–π* transition. The *trans* isomers also underwent a dramatic color change, from yellow to blue–green, upon addition of triflic acid, in agreement with the protonation of the azo moiety. When a slight amount of acid was added to the photogenerated *cis* isomers, an immediate *cis* to *trans* transformation was observed. This "proton-catalyzed" isomerization indicated that the *cis*-protonated form instantly produced the *trans*-protonated form which gave the *trans* isomer after the release of H$^+$. In this isomerization, it was presumed that protonation of the nitrogen atom changed its hybridization from sp^2 to sp^3, allowing a more facile rotation around the N–N bond (Scheme 9).

Nishihara and coworkers have recently investigated the photoisomerization behavior of platinum(II) complexes **10** and **11** containing an azobenzene on the metaldithiolene side and the bipyridine side, respectively (Scheme 10a) [38]. They also studied the photocontrolled multistability of a platinum complex **12** featuring two azobenzene moieties, one bounded to the dithiolato ligand and the other linked to the bipyridine ligand. All complexes feature strong absorption bands in the UV region, attributed to π–π* transitions of the azobenzene and bipy moieties. Complex **11** also displays at λ 405 nm a band that has been assigned to a MLCT-like transition, from the metalladithiolene (π) to the azobenzene (π*) level. In the lower

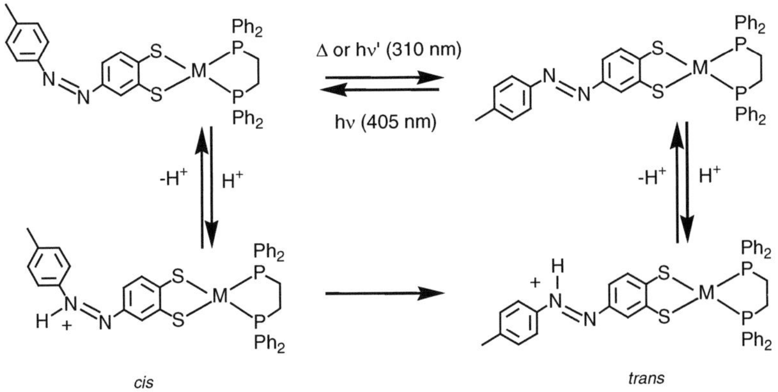

Scheme 9 Photo and proton responses of azobenzodithiolene metal complexes

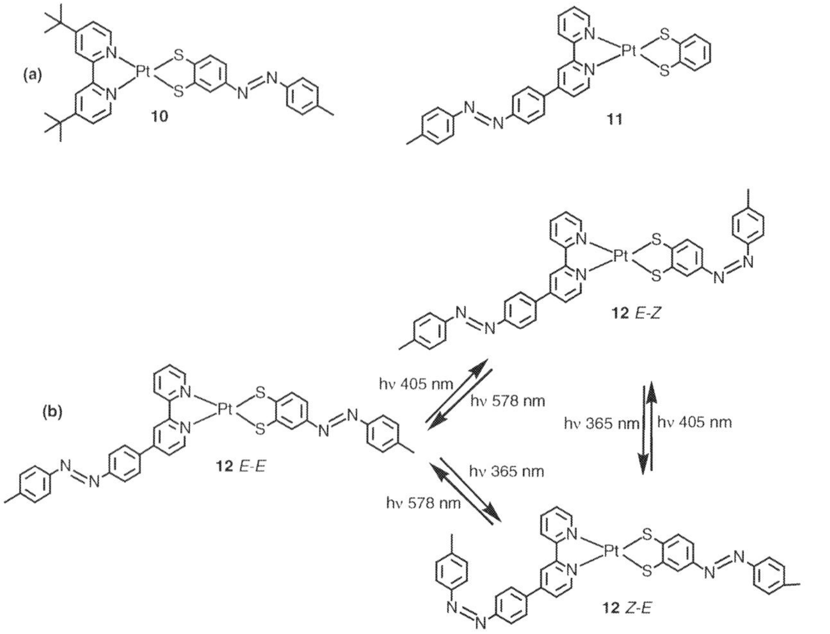

Scheme 10 (a) Chemical structures of complexes **10** and **11 b** Photoisomerization behavior of **12**

energy region, other bands were observed for both **10** and **11**, attributed as mixed-metal/ligand-to-ligand charge transfer (MMLL'CT) bands [39]. All these assignments were supported by TD-DFT calculations. These two complexes underwent both *trans/cis* photoisomerization, but at different wavelengths. For **10**, the *trans* to *cis* isomerization proceeded with a conversion of 45% in PSS, upon irradiation into the MLCT band at 405 nm. The back *cis* to *trans* isomerization occurred by excitation of the MMLL'CT band with 578 nm yellow light. On the other hand, complex **11** displayed *trans* to *cis* photoisomerization upon irradiation with 365 nm UV light (23%conversion in PSS), whereas the reverse isomerization was also observed by irradiation at 578 nm. Interestingly, complex **12** bearing an azobenzene on both the bipy and dithiolate ligands, showed photochromic behavior which was almost the superposition of those of complexes **10** and **11**. Thus, by using three monochromic lights, three out of the four possible states, the exception being the *cis,cis*-state, could be reversibly switched (Scheme 10b).

2.6 Azobenzene-Containing Metal Alkynyl Complexes

Organometallic alkynyl complexes exhibit a rich coordination chemistry with copper(I), silver(I), and gold(I) ions, and the ability of alkynyl groups to coordinate to metal centers in σ- and π-bonding modes has made them versatile ligands in the

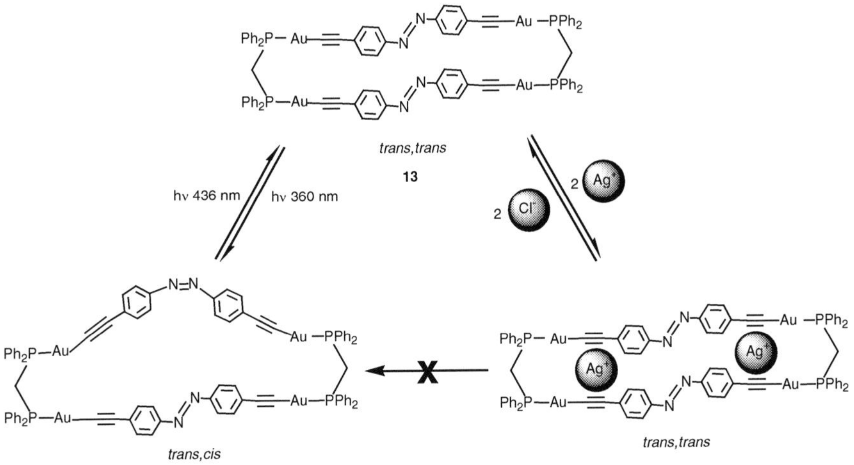

Scheme 11 "Locking" and "unlocking" mechanism upon addition and removal of Ag⁺ ions

synthesis of polynuclear metal complexes [40]. Very recently, Yam and coworkers prepared a tetranuclear gold(I) alkynyl phosphine complex **13** containing azobenzene functionalities and demonstrated the generation of a dual-input lockable molecular logic photoswitch [41]. Upon irradiation at 360 nm into the π–π* azo transition, *trans–cis* photoisomerization occurred, and it was estimated that only the *trans,cis* product was formed. The back *cis* to *trans* isomerization proceeded by excitation into the *n*–π* transition at 486 nm. Interestingly, introduction of Ag⁺ ions to **13** inhibited the photoisomerization process, probably because of π-coordination of adjacent pairs of alkynyls to the Ag⁺ ions which locks up the *trans* conformation of the azobenzene units. In contrast, upon abstraction of Ag⁺ with a chloride anion, the photoisomerization process could be restored, showing that photoswitching behavior can be controlled via silver(I) coordination/decoordination (Scheme 11).

A series of push–pull azobenzene containing ruthenium(II) σ-acetylide NLO-phores have also been prepared recently and their holographic properties, which are based on *trans–cis–trans* photoisomerization cycles of azobenzene, have been investigated in PMMA thin films [42, 43]. Surface relief gratings with good temporal stability were obtained upon short pulse laser irradiation ($\lambda = 532$ nm), showing that these organometallic photochromes are promising candidates for optical data storage applications.

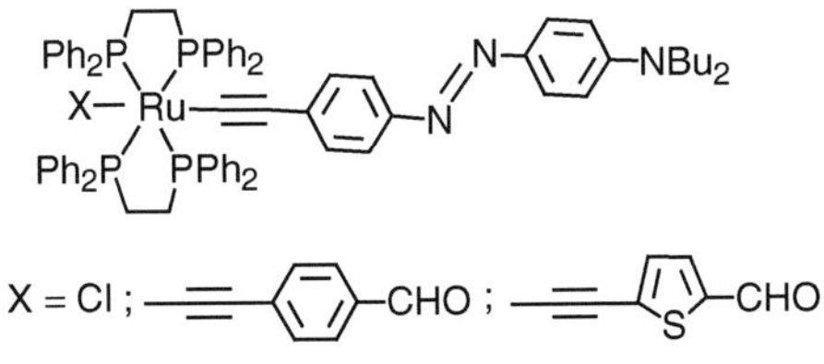

3 Metal Complexes Incorporating 1,2-Dithienylethene

3.1 Introduction

Diarylethene (DTE) derivatives that have been used as ligands for incorporation into transition metal complexes have recently received much attention. Coordination of DTE ligands opens up new perspectives for the design of photoswitchable molecules. 1,2-Dithienylcyclopentene and its perfluoro analog have particularly attracted increased interest for their excellent stability and fatigue-resistance properties. Irradiation at appropriate wavelength allows the interconversion of a nonconjugated colorless *open* form to the conjugated, colored, *closed* form [44, 45].

The open-ring isomer of DTE has parallel and antiparallel conformations which are in dynamic equilibrium. The closing process follows the rules according to Woodward–Hoffmann, in which the photocyclization occurs, via a conrotatory mechanism, only from an antiparallel conformation of the two thienyl rings. For unsubstituted DTE derivatives, the ratio of molecules in the parallel and antiparallel conformations is close to 1:1 and the cyclization quantum yield therefore cannot exceed 0.5. The photoreversion is significantly less efficient than the photocyclization, in most cases the quantum yield values typically not exceeding 0.1.

3.2 Organo-Boron DTE-Based Dithienylcyclopentene

3.2.1 Modulation of the Photochromic Properties by Ions

The special binding ability of the organo-boron functionalized 1,2-dithienylcyclopentene DTE-BMes$_2$ with fluoride (tetrammonium fluoride (TBAF)) and mercuric ions (Hg(ClO$_4$)$_2$) allows the modulation of its spectral properties, in its open and PSS states [46].

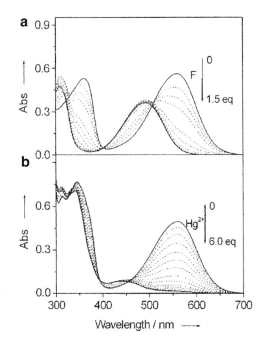

Upon addition of about 1.5 equivalents of TBAF, the absorption maximum of PSS (generated by irradiation with 365 nm) was blue-shifted from 560 to 490 nm. When about 6 equivalents of Hg(ClO$_4$)$_2$ was added, the absorption maximum of PSS was blue-shifted from 560 to 440 nm. Moreover, the absorption intensity decreased. The modulation mechanism is attributed to the Lewis acid–base interaction between a trivalent boron atom and a fluoride ion, and the complexation interaction between mercury and the sulfur atom (Fig. 2).

Fig. 2 Changes in UV–visible absorption spectra of closed-ring DTE–BMes$_2$ (PSS) (5.10^{-5} M) in THF solution upon addition of (a) 0–1.5 equivalents TBAF and (b) 0–6.0 equivalents of Hg(ClO$_4$)$_2$ (Reprinted with permission from [46])

Similar results were obtained in the case of bis(mesityl)boryl derivative DTE-(BMes₂)₂; its fluoride binding property is remarkably selective [47].

DTE-(BMes₂)₂

3.2.2 Up-Conversion Luminescent Switch Based on the Combination of DTE and Rare-Earth Nanophosphors

Up-conversion rare-earth nanophosphors (UCNPs) consisting of certain lanthanide dopants embedded in a crystalline host lattice can convert near infrared (NIR) excitation light into emission at visible wavelengths via the sequential absorption of two or more low energy photons. The hybrid system DTE–BMes₂/LaF₃:Yb, Ho loaded in PMMA film underwent a reversible photochromic reaction similar to those of the isolated species DTE–BMes₂ [48]. What is important is that DTE-BMes₂ has no absorption in the NIR region in both the open and PSS states, whereas LaF₃:Yb, Ho can emit visible luminescence by excitation at 980 nm, due to the large anti Stokes shift. A highly efficient hybrid nanosystem has been thus achieved via an intramolecular energy transfer process (Fig. 3).

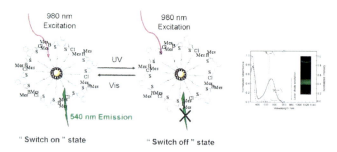

Fig. 3 UV–visible absorption spectra of DTE–BMes₂/LaF₃:Yb, Ho-loaded PMMA film before (*dashed line*) and after (*solid line*) irradiation with 365 nm light for 30 min, and the normalized up-conversion luminescence spectra of the prepared film (*dotted line*, $\lambda_{exc} = 980$ nm). *Inset* shows the image of the up-conversion emission of the film. (Reprinted with permission from [48])

3.2.3 Fluorescent Bodipy-Based Switches

Covalently linked BODIPY dyes (4,4-difluoro-4-bora-3a,4a-diaza-s-indacene) to the photochromic DTE unit allow the formation of new photoswitches that are highly emissive as open-ring isomers and in which the fluorescence is significantly quenched in closed-ring isomers [49]. Switching of the fluorescence is reversible and can be repeated (20 cycles) without significant loss of its intensity. Intramolecular energy transfer from the dye to the closed-ring form of the DTE is suggested to be the mechanism of fluorescence quenching.

3.3 Complexes Incorporating DTE-Based Pyridine, Cyano, or Carboxylate Ligands

3.3.1 Photochromism in Solution

The pyridine derivatives, the *monodentate* 1-(2-methyl-5-phenyl-3-thienyl)-2-(2-methyl-5-(4-pyridyl)-3-thienyl)perfluorocyclopentene and the *bidentate* 1,2-bis(2-methyl-5-(4-pyridyl)-3-thienyl)perfluorocyclopentene, incorporating the DTE moiety as the photochromic unit (4-py-DTE$_f$ [50, 51] and 4-py$_2$-DTE$_f$ [52]) and the nonfluorinated analog (4-py$_2$-DTE) [53] have been coordinated to various transition metals, allowing the change of physical properties of metal complexes.

4-py$_2$-DTE (X=N)
4-py-DTE (X=CH)

Branda showed that the ability of the pyridine ligand of the pyridinium salt of 4-py$_2$-DTE$_f$ to coordinate to a ruthenium center is modulated by interconverting the compound between its electronically insulated ring-open and electronically connected ring-closed form [54, 55].

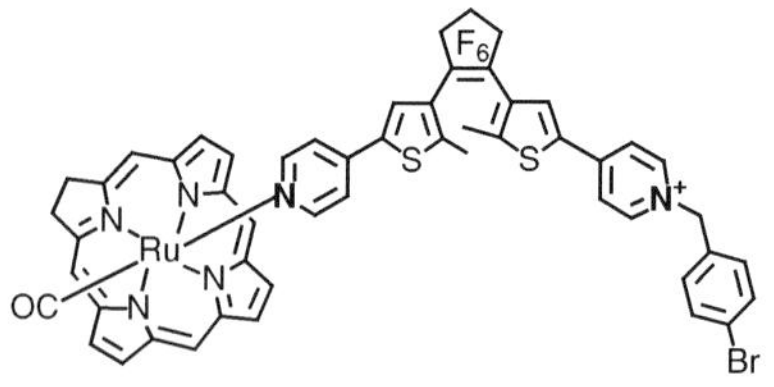

(Ru-TPP)(4-py-DTE_f_)

The *open-* and *closed*-ring monocationic DTE-based pyridine ligand exhibits different ability to coordinate the ruthenium porphyrin complex Ru(TTP)(CO)(EtOH) (TTP: *tetra*(4-methylphenylporphyrin)). The axial coordination of the closed-ring isomer of the pyridine complex which is electronically coupled with the pyridinium end group through to the π-conjugated DTE bridge is 1.5 times less effective than that of the related open isomer. This low selectivity is suggested to result from the fact that the free pyridine ring is not completely coplanar with the DTE framework.

Complexes of tungsten, rhenium, and ruthenium of 4-py-DTE$_f$ and 4-py$_2$-DT$_f$ have been developed by Lehn [56, 57]. These complexes exhibit good photochromic properties. A switching of fluorescence between their open and closed forms is observed when excited at 240 nm, a wavelength of irradiation that almost did not affect the state of the molecule.

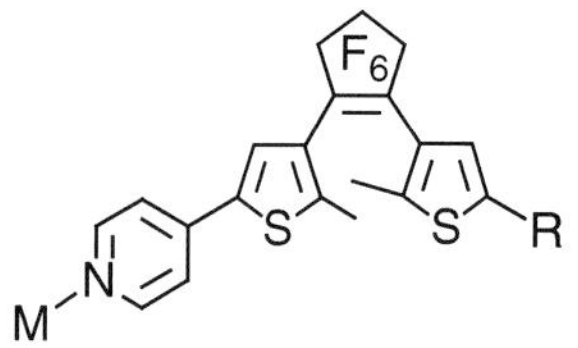

M = W(CO)$_5$, R = *N*-[W(CO)$_5$]pyridyl
M = W(CO)$_5$, R = *p*-phenol
M= [Re(bpy)(CO)$_3$(CF$_3$SO$_3$)],
R = *N*-[Re(bpy)(CO)$_3$(CF$_3$SO$_3$)]pyridyl
M= [Re(bpy)(CO)$_3$(CF$_3$SO$_3$)],
R =*p*-methoxy phenyl
M= [Ru(NH$_3$)$_5$](PF$_6$)$_2$, R = [Ru(NH$_3$)$_5$](PF$_6$)$_2$

The related metal complexes (W, Re, Ru) of the two cyano derivatives DTE–CN and DTE–C$_6$H$_4$–CN were found to be unstable.

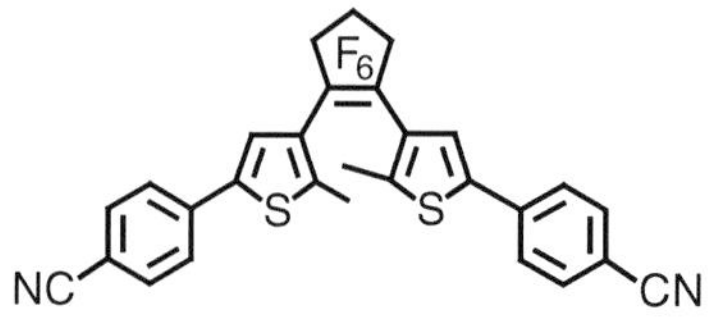

DTE-C_6_H_4_-CN

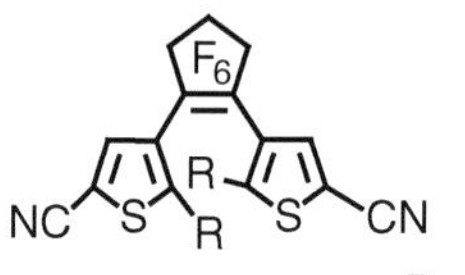

DTE-CN

R = C$_7$H$_{15}$

The photoresponsive ligand 4-py-DTE$_f$ has been combined with a spin-crossover complex, in order to achieve a molecular bistable spin system. The photoisomerization of the high spin Fe(II) complex Fe(4-py-DTE$_f$)$_4$(NCS)$_2$ allows the modulation of its magnetic properties [58].

3.3.2 Photochromism of Metal Complexes in the Single-Crystalline Phase

The photoreactivity of DTE derivatives in the crystalline state is of special interest because of their potential usefulness for holographic and three-dimensional memories. In crystals, molecules are regularly oriented and packed in fixed conformations. In many cases, free rotation is inhibited. Therefore, the photoreactivity in the crystalline phase is dependent on the space for free rotation of the thienyl rings and the conformation formed in the crystal lattice. Several reports on the synthesis of metal complexes of photochromic diarylethenes and their photo-reactivity in the single-crystalline phase or the photoswitching of the coordination structure have been reported. These studies demonstrate that complexation to metal ions does not prohibit the photochromic reactions of the diarylethene units in the single-crystalline phase.

Discrete Structure and Coordination Polymers

Metal complexes of monodentate (4-py-DTE$_f$) and bidentate (4-py$_2$-DTE$_f$) photochromic pyridine ligands and M(hfac)$_2$ (Zn(II), Cu(II), Mn(II)) (hfac = hexafluoroacetylacetone) have been synthesized [51]. The bidentate ligand 4-py$_2$-DTE$_f$ leads to the formation of coordination polymers (4-py$_2$-DTE$_f$)M(hfac)$_2$ whereas discrete 1:2 complexes were obtained for the monodentate ligand 4-py-DTE. The chain structure depends on the metal fragments, the complexes adopt a chain-like structure, a zigzag shaped in the case of (4-py$_2$-DTE$_f$)Cu(hfac)$_2$ and (4-py$_2$-DTE$_f$)Zn(hfac)$_2$ and an almost straight chain in the case of (4-py$_2$-DTE$_f$)Mn(hfac)$_2$ (Fig. 4). All photochromic units adopt an antiparallel conformation, in which photocyclization reaction takes place. This process has been monitored by means of polarized absorption spectroscopy.

The complexation of Cu(hfac)$_2$ and the isolated *closed-ring* isomers were performed to afford discrete 2:3 complexes closed-(4-py-DTE$_f$)$_2${Cu(hfac)$_2$}$_3$ and closed-(4-py$_2$-DTE$_f$)$_2$Cu(hfac)$_2$, probably as the result of reduced coordination ability of the pyridine ligand upon ring-closure. Photocyclization process induced a change in the coordination structure, as demonstrated by ESR studies in the case of (4-py$_2$-DTE$_f$)$_2$Cu(hfac)$_2$.

Similar results were obtained in the case of (4-py-DTE)ZnCl$_2$, two pyridines of two different molecules are coordinated to ZnCl$_2$ to form a linear coordination polymer [59]. It is noteworthy that, in this case, the complex adopts a parallel conformation and therefore no photochromism was observed in the crystalline phase.

Fig. 4 ORTEP drawings of X-ray crystallographic structures of linear chain complexes of the open-ring isomer (50% probability).
a (4-py$_2$-DTE$_f$)$_2$Zn(hfac)$_2$.
b (4-py$_2$-DTE$_f$)Mn(hfac)$_2$.
c (4-py$_2$-DTE$_f$) Cu(hfac)$_2$.
Hydrogen atoms are omitted for clarity. Only one repeating unit. (Reprinted with permission from [51])

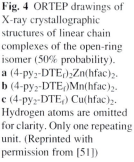

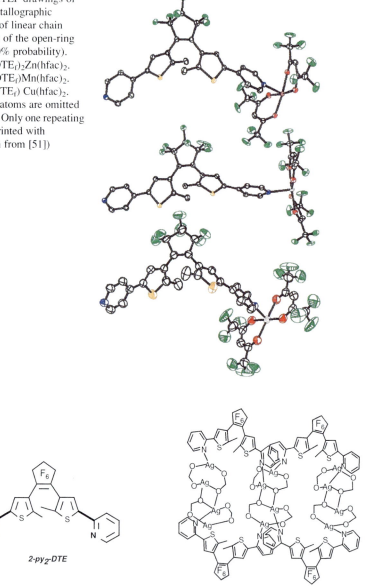

Fig. 5 One-dimensional double chain structure of [Ag$_2$(2-py$_2$-DTE)][(CF$_3$CO$_2$)]. (Reprinted with permission from [61])

The crystal of the discrete 2:1 complex (4-py$_2$-DTE)$_2$ZnCl$_2$ undergoes photochromic reaction by alternate irradiation with UV and visible light.

The structure and the photoreactivity depend on the substituents of the cyclopentene ring of the DTE unit, perfluorinated or not. Tian observed an enhancement of the photochromism, i.e., a higher quantum yield compared to that of the free ligand by coordination of the 1,2-bis(2-methyl-5(-4-pyridyl)-3-thienyl)*cyclopentene* to ZnCl$_2$ [53]. The structure of the coordination polymer, a zigzag chain, shows that the open-ring of the DTE adopts an antiparallel conformation in the crystal phase [60].

The related derivative 1,2-bis(2-methyl)-5-(**2**-pyridyl)-3-thienyl-perfluorocyclopentene (2-py$_2$-DTE) is able to coordinate Ag(I) ions [61]. Three novel Ag(I) complexes were prepared and all complexes display reversible photogenerated behavior in crystalline phase (Fig. 5).

1,2-Dicyano-1,2-bis(2,4,5-trimethyl-3-thienyl)ethene (*cis*-dbe)

The two cyano groups of 1,2-dicyano-1,2-bis(2,4,5-trimethyl-3-thienyl)ethene (*cis*-dbe) bridge two dimetal carboxylates of Mo(II) or Rh(II), to give a 1D zigzag infinite chain. However, the photochromic properties in crystalline phase are only observed for Mo complexes [62].

cis-dbe **trans-dbe**

Switch ON State   Switch OFF State

In the case of Cu(I) ions, the structure is composed of macrocations [Cu(*cis*-dbe)$_2$]$^+$ in which each metal center is coordinated with one CN group of the four DTE molecules [63]. Each *cis*-dbe bridges two copper(I) ions with two cyano groups, leading to an infinite network of metal cations. This copper coordination polymer shows reversible photochromism in the crystalline phase, whereas no modification is observed for the related [Cu(*trans*-dbe)(THF)(ClO$_4$)]$^+$ which forms a zigzag chain (Fig. 6).

Munakata also investigated the crystallographic structures of silver(I)-*cis*-dbe coordination polymers with five different anions (CF$_3$SO$_3$, C$_n$F$_{n+1}$CO$_2$; $n = 1$–4) which shows different photochromic reactions on the crystalline phase [64]. The differences are attributed to the varied Ag–S distances, controlled by the nature of the counter-anion (Fig. 7). These are the first examples of the metal ions coordinated to the thiophene rings of the DTE core, indicating that the soft Ag(I) ion has a relatively high affinity for sulfur donor atoms.

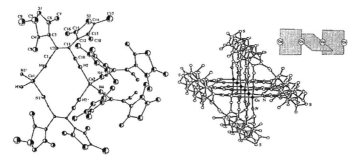

Fig. 6 a Partial molecular structure of [Cu(*cis*-dbe)₂][ClO₄]) showing the twisted ring-opened form of the DTE. **b** A view of the crystal packing showing a one-dimensional array of the metal ions. (Reprinted with permission from [63])

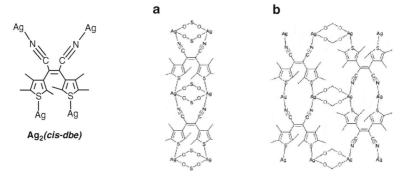

Fig. 7 Schematic views of (**a**) 1D structure of Ag₂(dbe)(CF₃SO₃)₂ and (**b**) 2D sheet structure of Ag₂(dbe)(X)₂ X = $C_nF_{n+2}CO_2$. (Reprinted with permission from [64])

1,2-Bis(2-methyl-5-carboxylic acid-3-thienyl)perfluorocyclopentene (DTE-5-CA)

1,2-Bis(2-methyl-5-carboxylic acid-3-thienyl)perfluorocyclopentene (DTE-5-CA), which was first synthesized by Branda [65] and incorporated into polymers, can also be used as a versatile ligand as demonstrated by Munakata [66].

DTE-5-CA

In the three complexes [Co(DTE-5-CA)(py)$_2$(MeOH)$_2$], [Cu(DTE-5-CA)(py)$_3$](EtOH)(py)$_{1.8}$, and [Zn(DTE-5-CA)(phen)(H$_2$O)], DTE-5-CA acts as a bis-monodentate bridging ligand through one of the oxygen atom of each carboxylate group to generate the 1D polymer. The two thienyl rings in all cases adopt antiparallel conformation and the distance between the two reactive carbon atoms are short enough to allow photocyclization in the crystalline phase. The coordination geometry and packing of the compounds significantly affect the photochromic performance.

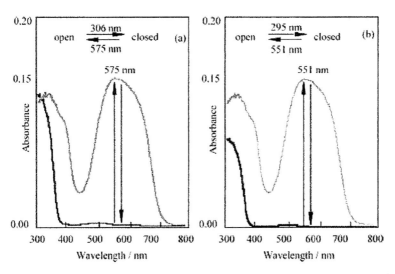

The Co(II) complex undergoes a solid-state structural transformation via the liberation of bound MeOH upon heating. Strikingly, the dynamic structural changes do not prohibit the reversible photoisomerization of the MeOH-desolvated form (Co–DTE–5-CA), indicating retention of the framework and excellent stability (Fig. 8).

Fig. 8 Absorption spectral changes of (**a**) Co–DTE–5-CA–MeOH and (**b**) Co–DTE–5-CA in the crystalline phase. (Reprinted with permission from [66])

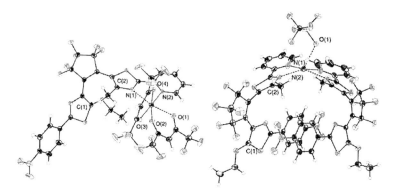

Fig. 9 ORTEP views of Cu(hfac)$_2$(dithiazole-py) (*left*) and Ag(dithiazole-py)$_2$CF$_3$SO$_3$ (*right*). (Reprinted with permission from [67])

Dithiazolylethene-based derivatives display an (*N,N*) chelating site from the pyridyl and thiazolyl fragments allowing the access to the monomeric metal complexes Cu(hfac)$_2$(dithiazole-py), Mn(hfac)$_2$(dithiazole-py), and Ag(dithiazole-py)$_2$CF$_3$SO$_3$ [67] (Fig. 9). None of these complexes display crystalline state photochromism.

3.4 Photoregulation of Luminescence

Among outputs, luminescence emission is considered to be one of the most attractive, owing to the ease of detection and the cheap fabrication of devices in which it is detected. Many examples of fluorescent photochromic molecules have been published, by combining a DTE unit with a fluorophore [4]. Incorporation of the DTE fragment into the ligands of transition-metal polypyridine complexes allows the photoreaction to proceed via a triplet state leading to a photoregulation of phosphorescence.

3.4.1 Complexes Incorporating DTE-Based Bipyridine Ligand

De Cola reported homo- and heterodinuclear systems in which the metallic fragments [Ru(bipy)$_3$]$^+$ and [Os(bipy)$_3$]$^+$ are bridged by a dithienylperfluorocyclopentene molecular switch, a phenylene group being used as a spacer [68–70].

[Structure diagram showing a dinuclear metal complex with M = Ru, Os, charge 4+, featuring bipyridyl ligands bridged by a dithienylethene unit with PF6]

In these metal systems, De Cola demonstrated that the photochromism can be extended from the UV region in the free ligand to the visible region corresponding to the MLCT transition. The extension of the excitation wavelengths to the visible region to trigger the photochromic reaction allows the use of less destructive visible light sources and is highly desirable. Direct evidence of triplet MLCT photosensibilization of the ring-closing reaction of diarylethene has been reported by means of transient absorption and time-resolved emission studies. The ring-closure reaction is in the nanosecond domain and the quantum yield decreases strongly in the presence of dioxygen. The photosensitization mechanism involving an internal conversion or intramolecular energy transfer from the ^{3}MLCT state to the ^{3}IL (DTE) state has been demonstrated. Once the ^{1}MLCT state is populated, intersystem crossing to the ^{3}MLCT state would occur leading to the sensitization of the photochromic reactive ^{3}IL(DTE) state that initiates the ring-closure reaction (Scheme 12). This behavior differs substantially from the photocyclization process of the nonemissive DTE free ligand, which occurs from the lowest ^{1}IL state on the picosecond timescale and is insensitive to oxygen quenching. Replacement of both Ru(II) centers by Os(II) completely prevents the photocyclization reaction upon light excitation into the lowest-lying ^{1}MLCT excited state.

A disadvantage of such system is that, in the trapping state, energy is transferred to the photochromic moiety that can consequently convert to the parent form. Therefore monitoring the occurrence or absence of energy transfer cannot be used as means for nondestructive readout, as the excitation affects the state of the system.

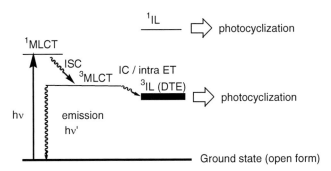

Scheme 12 Proposed qualitative energetic scheme for photosensitized photochromism by MLCT

The possible quenching by the sensitized photocyclization renders the photoluminescence from the ^{3}MLCT state less efficient. Upon conversion to their respective closed form, the ^{3}MLCT emission of both Ru(II) and Os(II) complexes are quenched. This is due to an energy transfer to the lowest excited state of the DTE unit that is lower in energy than those of the metal centers in the closed form.

The parent heterodinuclear Ru/Os complex was also investigated. The bridging unit in its open form allows an efficient energy transfer from the excited ruthenium to the acceptor osmium center. When the bridging DTE unit is in its closed form, the energy level IL(*closed*-DTE) drops down and is below of the energy level of the two metal centers, quenching both emissions.

3.4.2 Complexes Incorporating DTE-Based Terpyridine Ligand

The related dinuclear terpyridine complexes were investigated [71]. The Ru(II) and Fe(II) complexes, in their open forms, were found to be inert to UV irradiation but could be cyclized electrochemically as revealed by a cyclic voltammetry study. In contrast, the bis-Co(II) complex underwent efficient photochemical but not electrochemical cyclization. The corresponding Os(II) complex was neither photochromic nor electrochromic.

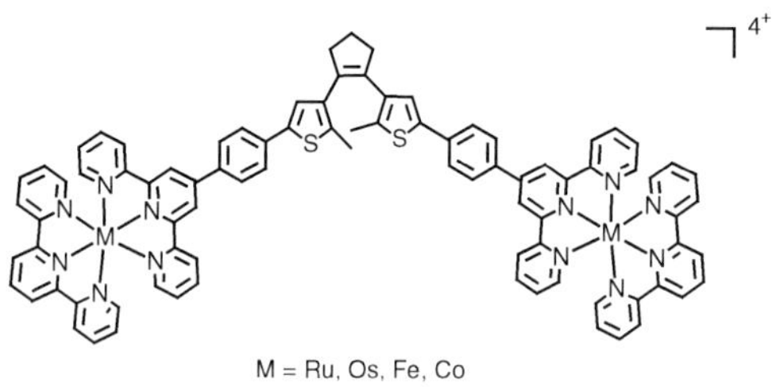

M = Ru, Os, Fe, Co

3.4.3 Complexes Incorporating DTE-Based Phenanthroline Ligand

Yam reported the synthesis and sensitized photochromic properties of a versatile diarylethene containing 1,10-phenanthroline ligand and their metal complexes (Re [72, 73], Pt [74]). Unlike other studies where the ligand is covalently connected to the DTE unit, phen1 displays an original design in which the ligand itself is part of the dithienyl framework.

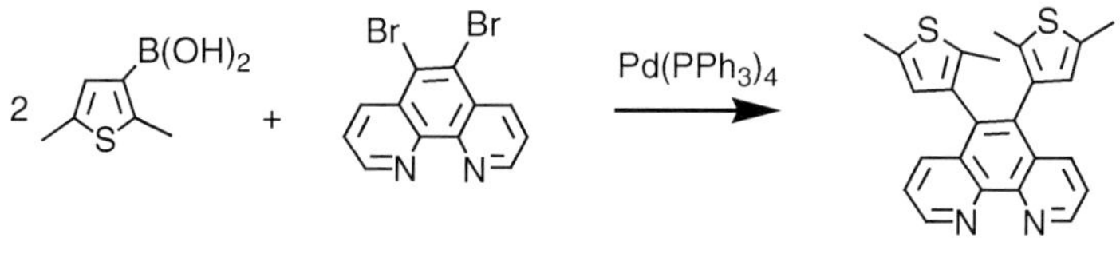

phen1

Unlike most other DTE systems, in which rapid interconversion of the two conformers results in a time-average ^{1}H NMR signals, the protons at the 4- and 7-positions of the sterically demanding phenanthroline moiety hinder the rotation of the thienyl rings. Thus, the parallel and antiparallel conformations of the free ligand phen1 and the rhenium complex [Re(CO)$_3$(phen1)(Cl)] (Re-phen1), are distinguished by the presence of two-well resolved sets of ^{1}H NMR signals. The structure of the free ligand phen1 shows that two thiophene rings tend to orient themselves perpendicular to the plane of the phenanthroline moiety.

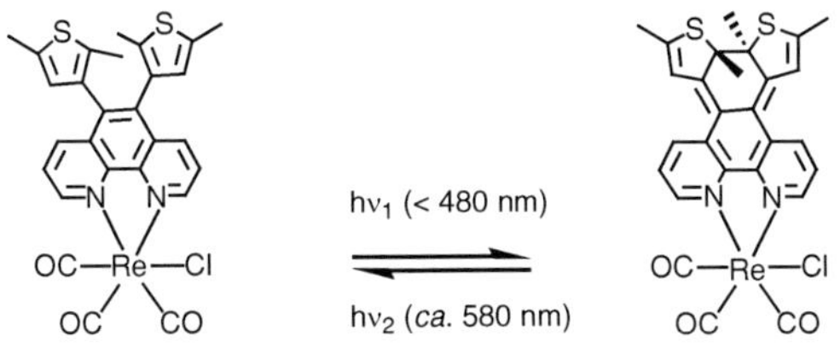

Re-phen1

The ligand phen1 has been also used to prepare the bis(alkynyl)platinum complex Pt-phen1. The structure of the two conformers, parallel and antiparallel, has been resolved. This is the first example of X-ray crystal structures in which both isomers of the same photochromic molecule have been determined.

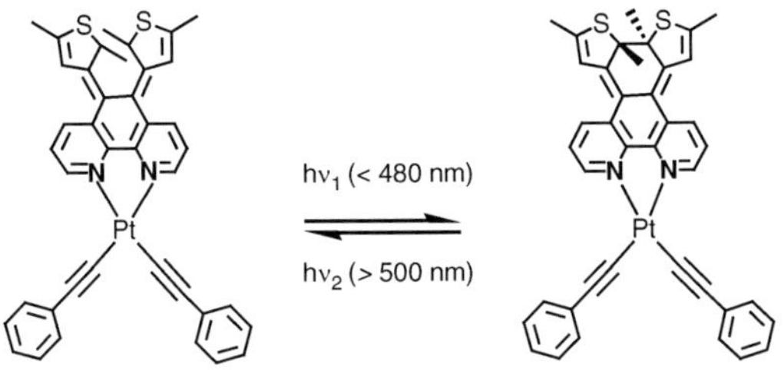

Pt-phen1

Perturbation of the photochromic and luminescence properties upon coordination to the metal center (Re, Pt) has been observed. A red shift of the absorption band for the closed isomer is attributed to the perturbation of the metal center in the complex. The emission of both phen1(*o*) and M-phen1(*o*) (MLCT) (*o*: open form, *c*: closed form) changes upon conversion to the closed form in the PSS. The strong red-shift of the emission observed for the closed form phen1(*c*) is attributed to the extension of the π-conjugation and has an IL (π→π*) phosphorescence in origin. A close resemblance of the emission of M-phen1(*c*) with that of phen1(*c*) suggests that the IL excited state, lower-lying in energy than that of the MLCT excited state, is the predominant emissive state.

Unlike the symmetrically substituted analogue phen1, the unsymmetrically substituted DTE containing 1,10-phenanthroline ligand phen2 and the corresponding Ru-phen2 exhibit no photochromic properties [75].

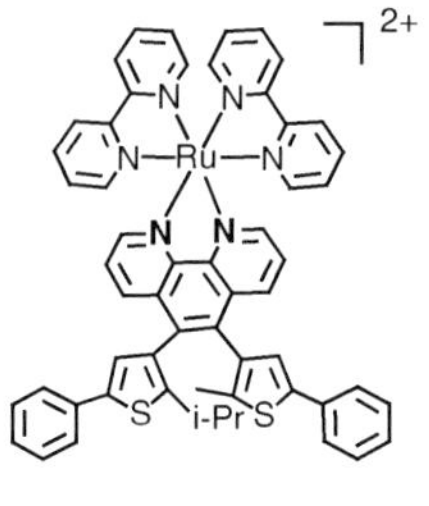

Ru-phen2

The diarylethene containing imidazo[4,5-f][1,10]phenanthroline has been synthesized but no examples of metal complexes are reported yet [76]. These ligands are sensitive to both light (UV/visible light irradiation) and chemical stimuli (alkali/acid treatment). A reversible four-state molecular switch has been realized by a single molecule.

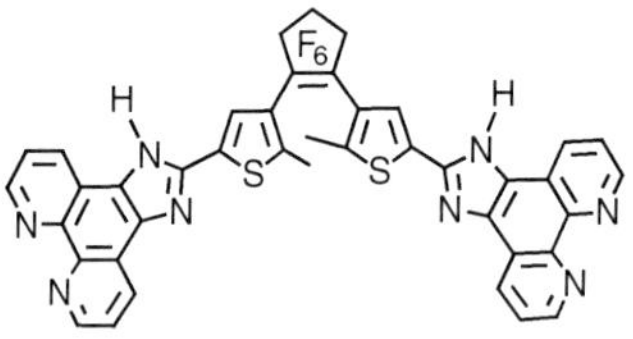

phen3 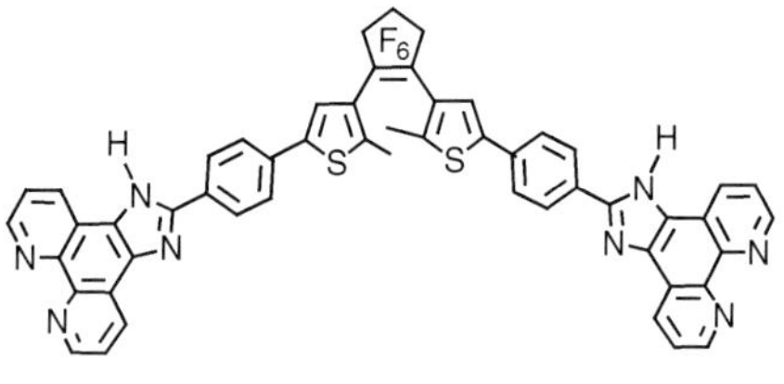   **phen4**

Maleimide Model

The study of a 1,2-bis(2-methylbenzothiophen-3-yl)maleimide model (phen-DAE) and two dyads in which the photochromic unit is coupled via a direct nitrogen–carbon bond (Ru–phen–DAE) or through an interverting methylene group (Ru–phen–CH$_2$–DAE) to a Ru–polypyridine chromophore has provided strong evidence for the participation of triplet state in the photochromic behavior of this class of diarylethenes [77]. Unlike previous studies, evidence of triplet reactivity case is obtained not only for the metal-containing systems but also for the isolated phen-DAE. A complete kinetic characterization has been obtained by ps–ns time-resolved spectroscopy. The experimental results are complemented by a combined ab initio and DFT computational study whereby the potential energy surfaces for ground state and lowest triplet state of the DAE are investigated along the reaction coordinate for photocyclization/cycloreversion.

Metal Complexes Featuring Photochromic Ligands

Ru-phen-DAE   Ru-phen-CH$_2$-DAE

3.4.4 Near-Infrared Photochromic Behavior (Imidazole–Pyridine)

Metal coordination of the DTE derivative and extension of the π-conjugated system through an enhancement of planarity provides an alternative and versatile route to a new class of photochromic compounds that show absorption and reactivity in the NIR region.

The Re(I) complexes featuring a DTE containing 1-aryl-2-(2-pyridyl)imidazole ligand has been demonstrated to exhibit NIR photochromic behavior, with a large red shift in absorption maxima upon photocyclization that has been brought by metal coordination-assisted planarization of the extended π-conjugated system [78].

λ_{max} 320 nm   580 nm   350 nm   710 nm

[Re] = Re(CO)$_3$Cl

3.5 Photoswitching of Second-Order NLO Activity

The NLO properties of metal-containing photochromic ligands have been evaluated by EFISH measurement for the open and PSS closed forms, and for the first time an efficient ON/OFF switching of the nonlinear optical (NLO) response was demonstrated [79]. Most molecules with large NLO activities comprise π-systems unsymmetrically end-capped with donor and acceptor moieties. In order to carry out the photoswitching of the NLO properties, a new type of 4,4′-bis(ethenyl)-2,2′-bipyridine ligands functionalized by a phenyl- and dimethylaminophenyl- DTE groups has been designed. These ligands have allowed the preparation of photochromic dipolar zinc(II) complexes. These molecules underwent an efficient reversible interconversion between a nonconjugated open form and a π-conjugated closed form when irradiated in the UV and visible spectral ranges, respectively.

D = H μβ₀ = 75 x 10⁻⁴⁸ esu
D = NMe₂ μβ₀ = 160 x 10⁻⁴⁸ esu

D = H μβ₀ = 1020 x 10⁻⁴⁸ esu
D = NMe₂ μβ₀ = 1800 x 10⁻⁴⁸ esu

3.6 Photoswitching of a Magnetic Interaction

The first example of a magnetic metal–radical interaction was achieved by the coordination of a DTE (1,2-bis(2-methyl-benzothienyl-3-yl)perfluorocyclopentene) based-1,10-phenanthroline ligand containing a nitronyl nitroxide radical with a Cu(II) ion [80]. Mixing this ligand with [Cu(hfac)₂] in toluene led to a hypsochromic shift of the absorption maxima of the closed-ring isomer due to complexation. ESR measurement in toluene of the open-ring isomer of the Cu complex gave a spectrum that is the superposition of the spectra from the nitroxide radical and Cu(II). Photoirradiation gives rise to a new peak due to a large exchange interaction; the exchange interaction difference between open- and closed- isomers was estimated by ESR spectral change to be more than 160-fold (Fig. 10).

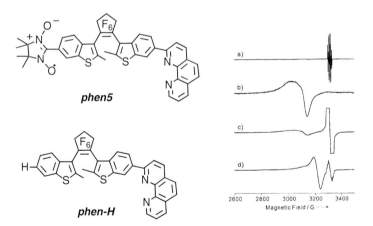

Fig. 10 X-band ESR spectra in toluene solution at room temperature (9.33 GHz, 1 mW, 2,600–3,500 G region). **a** Spectrum of phen5. **b** Spectrum of Cu(hfac)₂(phen-H). **c** Spectrum of Cu(hfac)₂ (phen5). **d** Spectrum of closed-Cu(hfac)₂(phen5); (in the photostationary state under irradiation with 366 nm light). The spectra were obtained with 0.5 G modulation amplitude for **a**, and with 32 G modulation amplitude for **b**–**d**. The same sample was used for measuring spectra **c** and **d**. (Reprinted with permission from [80])

3.7 Photo- and Electrochromic Properties of Metal-Based DTE Derivatives

Electron transfer is an appealing process allowing the same transformation as in photochemistry, i.e., cyclization-reopening. In this context, the DTE unit has proven to be useful for the elaboration of photo and/or electrochromic metal-based systems specially devoted to the control of the electronic communication in bimetallic complexes. Organometallic complexes consisting of π-conjugated system bridging two redox active metal centers have proven to be efficient molecular wires (Fig. 11).

The attachment of the iron organometallic moieties [(η^5-C$_5$Me$_5$)Fe(dppe)] (dppe = diphenylphosphinoethane) at the 5 and 5′-positions of the thiophene rings, using a C≡C as a linker group, improves the efficiency of the photochromic process, compared to that of the free alkyne which is unchanged even after prolonged UV irradiation. The electronic communication performance between the two metal centres could then be switched ON and OFF, as illustrated by the large difference of the K_C (conproportionation constant) values of the two open and closed forms (K_C(ON)/K_C(OFF) = 39).

M⎯≡⎯[thiophene]⎯[thiophene]⎯≡⎯M

M = (η^5-C$_5$Me$_5$)Fe(dppe) [81]
    Ru(dppe)$_2$(Cl) [82]

The utilization of the ruthenium carbon-rich fragments [Ru(dppe)$_2$(Cl)(C≡C)] gives rise to a light- and electrotriggered switch featuring multicolor electrochromism [82]. Quantitative cyclization occurs upon oxidation at remarkably low potential, far below pure DTE processes that generally occur around 1 V. This is the result of the unique electronic structure of the ruthenium complexes that leads, in radical species, to electronic delocalization on the carbon-rich ligand including the thiophene rings, allowing a radical coupling in the open state to form the more stable closed-ring isomer.

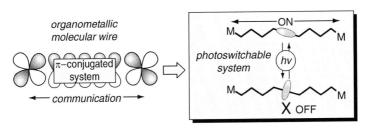

Fig. 11 Organometallic photoswitchable molecular wires. (Reprinted with permission from [81])

Extension of this work by directly σ-bonding the DTE unit with the redox active metal fragments M(η^5-C$_5$H$_5$)L$_2$ (M = Fe, Ru; L$_2$ = (CO)$_2$, (CO)(PPh$_3$), dppe) leads to bimodally stimuli-responsive, photo- and pseudoelectrochromic behavior with the remarkable switching factor K_C(closed)/K_C(open) up to 5.4 × 10^3 [83].

M = (η^5-C$_5$H$_4$)Fe/Ru(CO)$_2$

(η^5-C$_5$H$_4$)Fe/Ru(CO)(PPh$_3$)

(η^5-C$_5$Me$_4$)Fe(dppe)

It is noteworthy that the dicationic closed form C'^{2+} is diamagnetic, as a result of the coupling of the radical centers. Its carbene structure has been established on the basis of NMR data and the crystallographic structures of isolated iron C'^{2+} species.

The "electrical communication" between metallic sites could be probed by the presence or absence of the intervalence transition when the mixed valence state of the system is formed. Coudret and Launay have demonstrated the ON/OFF switch of the intervalence transition of the dinuclear cyclometallated Ru complex; the mixed valence state is, however, unstable [84].

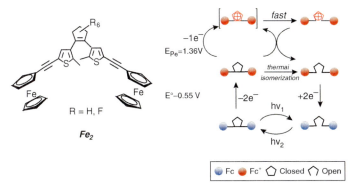

Fig. 12 Proposed mechanism for the perfluorinated bimetallic Fe$_2$ system. (Reprinted with permission from [85])

Later on, they investigated the bis(iron) derivative Fe$_2$ and showed that the presence of the ferrocenyl substituent makes possible a quantitative thermal reopening upon partial or complete oxidation of the redox active system without oxidizing the photochromic core [85, 86]. This process is catalytic in electrons in the case of the perhydro-DTE compound. The thermal reopening of neutral DTE unit is facilitated by electron-withdrawing substituent as *gem*-dicyanovinyl (Fig. 12).

3.8 Other DTE-Based Metal Complexes

The X-ray crystallographic structures of a series of DTE–organoruthenium complexes, DTE–(RRuL$_m$)$_n$ (RRuL$_m$ = (η^6-C$_6$H$_5$)Ru(η^5-C$_5$Me$_5$): n = 1,2; (η^6-C$_6$H$_5$) RuCl$_2$(PPh$_3$): n = 1, 2; (η^5-C$_5$Me$_4$)Ru(CO)$_2$: n = 1, 2) reveal the antiparallel conformation of the two thiophene rings suitable for the photochemical ring-closure. The efficiency of the photochromic process of these DTE–Ru complexes depends on the nature of attached metal fragments. However, no significant photochromic behavior is noticed for the benzofused derivative, and UV irradiation of the arene complexes of (η^6-C$_6$H$_5$)RuCl$_2$(PPh$_3$) induces the irreversible dissociation of the arene ligand [87].

The DTE decorated with phosphine ligands has been coordinated to gold. The electronic differences between the two isomers of the free bis(phosphine) ligand was demonstrated by ^{31}P NMR spectroscopy [88].

3.9 DTE-Based Ligands in Catalysis

Although examples of regulation of catalysis by azobenzene derivatives has been reported – a concept which relies on the changes in molecular geometry that result from the light-induced *cis–trans* isomerization – their successes are limited by the thermal reversibility of the azobenzene derivatives and cannot be used for practical applications. The approach developed by Branda takes advantages of the thermal stability of the DTE derivatives and harness the differences in geometry to regulate metal-catalyzed reactions [89].

The copper(I)-catalyzed cyclopropanation of styrene with ethyldiazoacetate was used as a model reaction.

oxazoline rings at the C-2 position

DTE-5-oxazoline

oxazoline rings at the C-5 position

When the oxazoline ring is located at position 2 of the two thienyl rings of the DTE, optically pure binuclear helicate, composed of two DTE-based ligands wrapped around two copper centers, is formed. Since this open-ring form is flexible, the isomeric states of compounds do not affect the stereoselectivity of the cyclopropanation when they are used as ligands and neither the enantioselectivity nor the diastereoselectivity is significantly altered.

Ring-open 5-oxazoline DTE	30% ee	50% ee
Ring-closed 5-oxazoline DTE	<5% ee	<5% ee

Changing the location of the oxazoline units from the "external" position to the internal position (position 5) allows the formation of a monomeric species in which the metal is in a C_2-chiral environment. Measurable but low enantioselectivities were observed by using the open-ring form whereas the closed-ring counterpart, isolated after UV irradiation, did not lead to significant stereoselectivity as a result of a more rigid architecture which prevents a suitable coordination site for the metal center.

3.10 Multi-DTE Metal Complexes

A few examples of metal complexes containing several DTE units have been reported. Generally, like their organic congeners [90–92], the conversion to the fully closed isomer is not observed; this is attributed to intramolecular transfer from the reactive state (open part) to the closed-ring DTE part, the lower-lying excited state.

Phosphorescent heteroleptic Ir(III)-based switches were designed and synthesized; an additional enhancement of the phosphorescence modulation has been observed by incorporating two accepting DTE units [93].

3.10.1 DTE-Based Phenolic Ligands (Oxy Anionic Ligands)

The preferential coordination of the tin(IV) porphyrins to oxyanionic ligands has been used for the elaboration of the photochromic fluorophore Sn(TTP) (DTE–Ph–O)$_2$ (TTP: 5,10,15,20-tetratolylporphyrinato) in which two phenolic derivatives of DTE (DTE–Ph–O) are axially coordinated in *trans* position. Small changes of the fluorescence intensity in the luminescence modulation are observed in the PSS [94].

A Pt macrocycle containing four DTE units has been described. Upon UV irradiation, two photo-induced cyclization reactions occur. The converted closed isomer Pt(*c,o,c,o*) has been isolated and characterized by ^{1}H NMR spectroscopy. The high value of the cyclization quantum yield (0.64) is attributed to the enforced antiparallel conformation in the macrocycle [95].

4 Photochromic Spiropyran and Spirooxazine-Containing Metal Complexes

4.1 Introduction

SP/SO are one of the most attractive classes of photochromic molecules and have been extensively studied due to their fatigue resistance and good photostability [96, 97]. The photochromic properties of SP were first recognized by Fischer and Hirshberg in 1952 [98]. The photochromism is attributed to the photochemical cleavage of the spiro C–O bond, forcing the molecule to open up. This ring-opening reaction results in the extension of the π-conjugation in the colored photomerocyanine (MC) form which is thermally unstable and readily reverted by thermal reaction or by absorption of visible-light irradiation (Scheme 13). Numerous studies on the nature of intermediates have been developed, including transient spectroscopy on the nanosecond timescale.

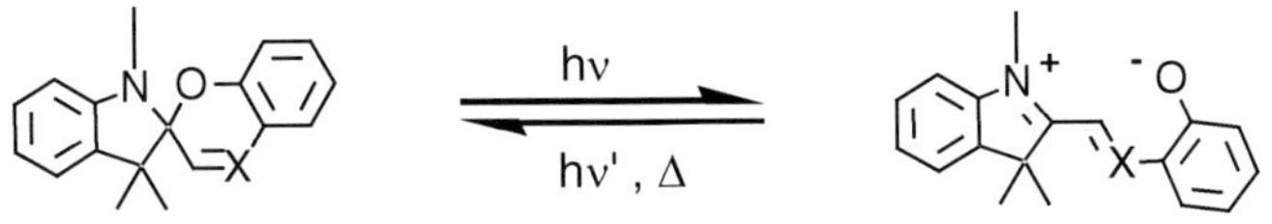

X = CH : *SP*
X = N : *SO*

Scheme 13 Photochromic interconversion of a typical spiropyran and spirooxazine

4.2 Ferrocenylspiropyran

Nishihara reported the synthesis of a ferrocenylspiropyran compound, and demonstrated that the thermodynamic stability of the MC form can be completely and reversibly switched by a combination of SP/MC photoisomerization and Fc/Fc$^+$ redox cycle [99]. Irradiation of a pale yellow solution of Fc–SP in methanol with UV light (365 nm) afforded a red purple solution ($\lambda_{max} = 530$ nm) ascribed to the open MC form. The MC molar ratio in the PSS was estimated to be 75% in methanol. The λ_{max}, yield in the PSS and thermodynamic stability were also found to depend on the polarity of the solvent: in dichloromethane the open form ($\lambda_{max} = 578$ nm) was formed in only 56% yield in the PSS, and isomerized to Fc–SP much faster than in methanol at 20°C. Upon chemical one-electron oxidation to the resulting Fc$^+$–SP, two new absorption bands appeared at 515 and 1,250 nm. This ferrocenium complex showed complete SP to MC photoisomerization with UV light irradiation and, unlike Fc–MC which readily isomerized to Fc–SP within few hours at 20 °C or upon visible irradiation, Fc$^+$–MC was stable towards heat and light. The reason for the MC stabilization by Fc$^+$ was not fully clarified, but the large blue shift for Fc$^+$–MC relative to Fc–MC suggested a strong electronic effect of Fc$^+$ as a result of π-conjugation. The reversible photoisomerization and redox cycle was also studied in a polymer matrix and showed the same behavior as that in solution.

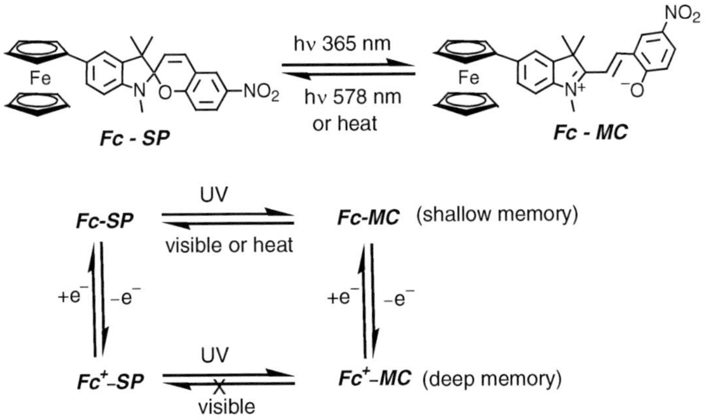

4.3 Porphyrin Spiropyran Metal Complexes

Bahr et al. have developed a new dyad by linking a photochromic nitrospiropyran moiety to a zinc (Porph$_{Zn}$–SP) porphyrin [100]. They showed typical photochromic behavior, i.e., conversion to the open merocyanine form upon UV excitation which in turn closes to the spiro form thermally or by irradiation into its visible absorption band. The emission properties of the two forms, Porph$_{Zn}$–SP and Porph$_{Zn}$–MC, were investigated. The fluorescence of the porphyrin is unperturbed by the attached SP moiety in the closed form, whereas the porphyrin first excited singlet state is quenched by the merocyanine with a quantum yield of 0.93, reducing the lifetime from 1.8 ns to 130 ps. The quenching of luminescence was assigned to a singlet–singlet energy transfer. Thus, this photoswitchable quenching phenomenon provides light-activated control of the porphyrin first excited states.

4.4 Spiropyran- and Spirooxazine- Containing Polypyridine Metal Complexes

SO are structurally and functionally very similar to SP. A series of rhenium complexes featuring a spironaphthoxazine-based pyridine ligand (SNOpy) [Re(CO)$_3$(phen)(SNOpy)]$^+$, [Re(CO)$_3$(4,4'-Me$_2$-bipy)(SNOpy)]$^+$, and [Re(CO)$_3$(4,4'-^{t}Bu$_2$-bipy)

(SNOpy)]⁺ (SNOpy: 1,3,3-trimethylspiroindoline naphthoxazine-9′-ylnicotinate) have been synthesized and their photochromic behavior demonstrated [101–103].

The Re–SNOpy complexes exhibit strong luminescence; the two emission bands are attributed to the ligand-centered (LC) fluorescence and phosphorescence. An assignment as MLCT phosphorescence was ruled out, since the emission energies are insensitive to the diimine ligands, irrespective of their different π* orbital energies. The formation of the MC (Re–MCpy) open form upon MLCT excitation is suggestive of an efficient photosensitization of Re–SNOpy by the MLCT excited state. It is likely that an intramolecular energy transfer from ³MLCT to the SNOpy moiety occurs, to give the ³SNOpy state, which would either return back to the ground state with emission of light or undergo the ring-opening process to give the colored Re–MCpy form.

The Re–MCpy form is thermally unstable and readily undergoes thermal bleaching which follows first-order kinetics to the closed form. The decoloration rate constant of the Re–MCpy form strongly depends on temperature and on the nature of the solvent used, the zwitterionic Re–MCpy form being more stabilized in polar solvent such as MeOH. This constant decreases, to a small extent, upon addition of $ZnCl_2$; this could be attributed to the stabilization of the Re–MCpy form by the formation a Zn^{2+} complex [101].

The related 2,2′-bipyridine derivatives have been prepared and incorporated in Re complexes [Re(CO)₃(Cl)(N,N-SNObipy)] [102]. Their X-ray crystal structure reveals an orthogonal arrangement of the indoline and naphthoxazine planes (interplanar angle 88°2″) and a relatively longer spiro C–O bond (1.47 Å), these features are commonly observed in related systems. Unlike the Re–SNOpy complexes [Re(bipy)(CO)₃(SNOpy)]⁺, Re–SNObipy complexes do not exhibit any photochromism with MLCT excitation. This is probably because the energies of the ³MLCT excited state of the complexes estimated from the phosphorescence of the complexes (~166–188 kJ mol⁻¹) are insufficient for the sensitization when compared to the triplet excitation states of spironaphthoxazines (~210–225 kJ mol⁻¹). The switching of their emission properties (³MLCT to ³LC phosphorescence of the merocyanine moiety) upon conversion to the open form has been reported.

Fig. 13 Time-dependent UV–visible absorption spectral changes in the open form of Re(CO)₃(Cl) (SObpy4) in acetonitrile after excitation at 365 nm. The *inserts* show the overlaid UV–visible absorption spectra at different decay times and the decay trace at the absorption maximum at 604 nm with time. (Reprinted with permission from [102])

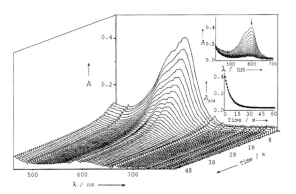

The kinetics for the bleaching process (Re–SNObipy to Re–MCbipy) have been determined for both the free ligands and the complexes (Fig. 13).

By incorporating these SNObipy ligands into various metal complex systems, their photochromic properties can be tuned and perturbed without the need for tedious synthetic procedures for the organic framework. As an extension of her work, Yam investigated a series of substituted spironaphtoxazine containing 2,2′-bipyridine and their Zn complexes bis-thiolate [103]. Upon excitation at 330 nm, all the ligands and complexes exhibit photochromic behavior.

SP or SO units have also been bounded to 2,2′-bipyridine and 1,10-phenanthroline: (1) the bipyridine moiety has been attached either to the pyran part (SPbipy1) or the indoline part (SPbipy2) of the SP skeleton, (2) the bipyridine ligand SPbipy3 bears a bipyridine at each end of the SP skeleton, (3) a phenanthroline-based SO attached to a bipyridine ligand (bipySOphen) [104].

The photophysical, photochemical, and redox properties of mononuclear complexes [M(bpy)$_2$(*NO$_2$-SPbipy*)]$^{2+}$ (M = Ru, Os) have been investigated [105]. These metallated nitrospiropyran compounds undergo efficient electrochemically induced conversion to the merocyanine open form, by first reducing the closed form and subsequently reoxidizing the corresponding radical anion in two well-resolved anodic steps. Metal complexation of the SP results in a strongly decreased efficiency of the ring-opening process as a result of energy transfer from the reactive excited SP to the MLCT excited state. The lowest excited triplet state of the SP in its open MC form is lower in energy than the excited triplet MLCT level of the [Ru(bpy)$_3$]$^+$ moiety but higher in energy than for [Os(bpy)$_3$]$^+$, resulting in energy transfer from the excited ruthenium center to the SP but inversely in the osmium case.

The heterobinuclear complex *Ru–SP–Os* has been synthesized by using the "chemistry-on-the complex" strategy [70]. To the Ru mononuclear complex a free chelating site was introduced, to which the Os moiety was subsequently connected. However, this dinuclear complex was shown to be inactive; no conversion to the open form occurs whatever the irradiation wavelength used (UV or visible (450 nm) light).

Ru-SP-Os

A dramatic increase in the ring-closing rate was observed for the *Ru–R–SOphen* (R = H, C$_{16}$H$_{33}$) complexes, the effect being attributed to the strong electron-donation (d_πRu → π*(*N,N*)) of the organometallic fragment [Ru(bipy)$_2$]$^{2+}$ [106].

R = H, OC$_{16}$H$_{33}$

Ru-R-S Ophen

A series of M^{II}tris(spiro[indolinephenanthrolinoxazine]) (Mn, Fe, Co, Ni, Cu, Zn) have been synthesized, resulting in tunable and significantly increased photoresponsivities (photocolorabilities) [107]. A significant stabilization of the MC form results from metal complexation. An increase in charge density in the oxazine of the molecule leads to destabilization of the MC form Metal complexation is expected to decrease the negative charge density of the oxazine moiety through inductive effects and to increase the charge density through $d_\pi M \rightarrow \pi^*(N,N)$ donation. The increase in MC stability suggests that inductive effects play a dominant role.

M-SOphen   M = Zn, Fe, Co, Mn, Ni, Cu

4.5 Diastereomeric Isomerism in $[(\eta^6\text{-spirobenzopyran})Ru(C_5Me_5)]^+$

η^6-Coordinated spirobenzopyran complexes are known for ruthenium and chromium. Ruthenium–arene complexes $[(C_5Me_5)Ru(\eta^6\text{-SP-X})]^+$ (SP–X benzopyran) are formed as two diastereoisomers where complexation occurs in the indoline (A) or dihydrobenzopyran part (B) of the SP–X ligand [108]. UV irradiation of A does not cause any color change, but induces an inversion of the configuration of the chiral spiro carbon atom (A′).

A   A′   B

$(\eta^6\text{-SP})Cr(CO)_3$ complexes (A ring adducts) were studied, the ring formation during their synthesis and ring-closing reaction during and after UV irradiation proceeding diastereoselectively to let chromium atom and oxygen atom of the pyran part located on the same side of the indoline ring [109].

4.6 Complexes of Spiropyrans with Metal Ions

Intramolecular bidentate metal ion chelation (Ca^{2+}, Zn^{2+}) gives rise to thermally stable SP/MC photoswitches [110, 111]. The addition of $Zn(ClO)_4$ and $Cu(ClO_4)$

gives rise to the stabilization of two geometric isomers, *cis*-MC and *trans*-MC, via the coordination to the COOH coordinating group attached to the indolino ring. This is the first evidence of the *cis*-MC isomeric form being observable by UV–visible spectrophotometry over a significant period of time [112].

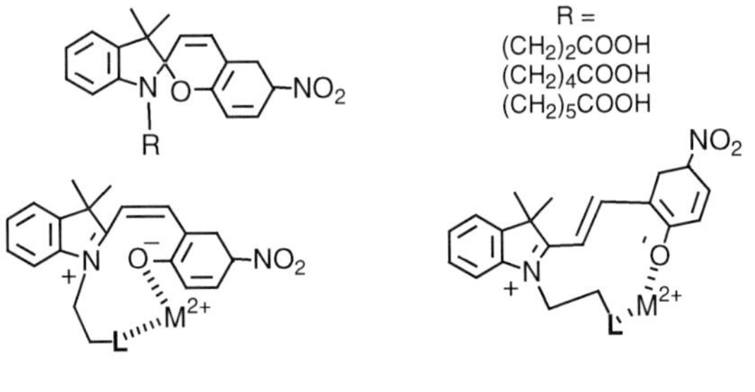

R =
(CH$_2$)$_2$COOH
(CH$_2$)$_4$COOH
(CH$_2$)$_5$COOH

cis-MC   **trans-MC**

5 Photochromic Metallocenyl Benzopyran Derivatives

Like SP, benzopyrans are converted under UV irradiation into colored merocyanine forms. The reaction is reversible and the back closure reaction takes place either thermally or upon irradiation in the visible range.

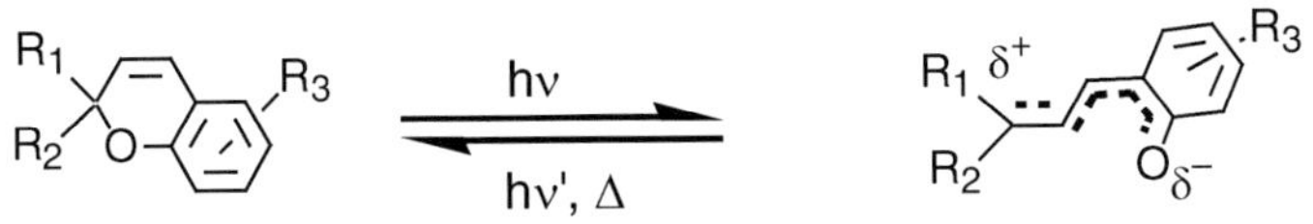

A series of ferrocenyl compounds such as **14** have been prepared and their photochromic properties investigated [113–115]. These studies have shown that the introduction of a ferrocenyl group in the 2-position modifies the photochromic behavior of these compounds, i.e., an extended wavelength range with two absorption bands around 450 and 600 nm for the open form, an increased of the closure kinetic constants and a good resistance to fatigue. More recently, the synthesis and photochromic properties of the corresponding ruthenocenyl **15** and osmocenyl **16** derivatives have been reported. These complexes showed only one absorption band in the visible region near 500 nm, but again an increase of the

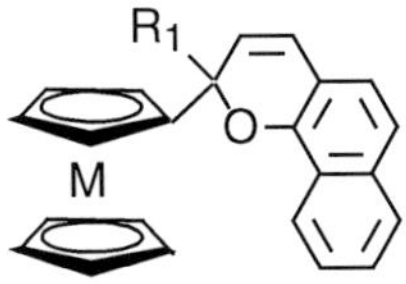

M = Fe: **14**; Ru : **15**; Os : **16**
R$_1$ = Me, Ph

bleaching kinetics, when compared to the parent alkyl- and phenyl-substituted benzopyrans.

6 Dimethyldihydropyrene Metal Complexes

Among the diarylethene family, the dimethyldihydropyrene (DHP)/cyclophanediene (CPD) system is interesting because it is a rare example of negative photochrome, where the thermally stable DHP closed form is colored. Upon irradiation with visible light, DHP is converted to the colorless, open CPD form, and the reverse reaction occurs either photochemically with UV light or thermally [116, 117].

Only a few examples of organometallic complexes featuring DHP derivatives are known. Mitchell and coworkers have recently reported the synthesis and photochromic properties of benzodimethyldihydropyrene chromium tricarbonyl and ruthenium cyclopentadienyl complexes, in which the Cr(CO)$_3$ and Ru(C$_5$H$_5$)$^+$ fragments are coordinated to the fused benzene ring [118, 119]. The photochromic behavior of these compounds were studied and found to depend on the nature of the organometallic moieties. Surprisingly, the chromium complex did not show any photoisomerization at all. Conversely, the ruthenium complex was found to photo-open quantitatively, but at 30% of the rate of the uncomplexed parent compound. Interestingly, photo-closing of the CPD to the DHP form occurred either photochemically with UV light, electrochemically on reduction, or thermally.

More recently, Mitchell, Berg and coworkers prepared divalent ytterbium and iron metallocenes based on cyclopentadienyl-fused DHP [120]. These complexes were isolated as a mixture of *rac* and *meso* isomers in 3:2 and 1:1 ratio for Fe and Yb, respectively. Their photochemical reactivities were also investigated, and photolysis in the visible at λ >490 nm did not result in isomerization to the CPD opened forms, in contrast to the lithium salt LiCpDHP [121].

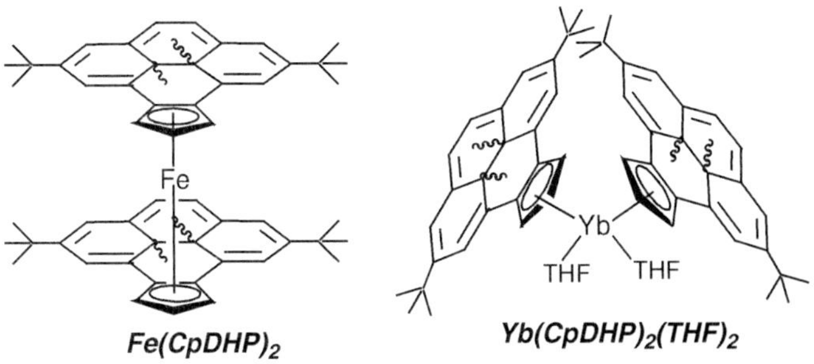

Fe(CpDHP)₂

Yb(CpDHP)₂(THF)₂

Very recently, Nishihara and coworkers prepared a bis (ferrocenyl)dimethyldihydropyrene (*Fc₂DHP*) in which DHP and metal complexes are connected through an ethynyl moiety [122]. This complex exhibits both reversible DHP/CPD photochromic behavior and photoswitching of the electronic communication the two ferrocene sites. In addition, Fc₂CPD also showed a redox-assisted closing reaction of the photogenerated CPD form with oxidation of the ferrocene moieties. This system is unique since the ring-closing is triggered by oxidation of not the DHP unit but the ferrocene moiety.

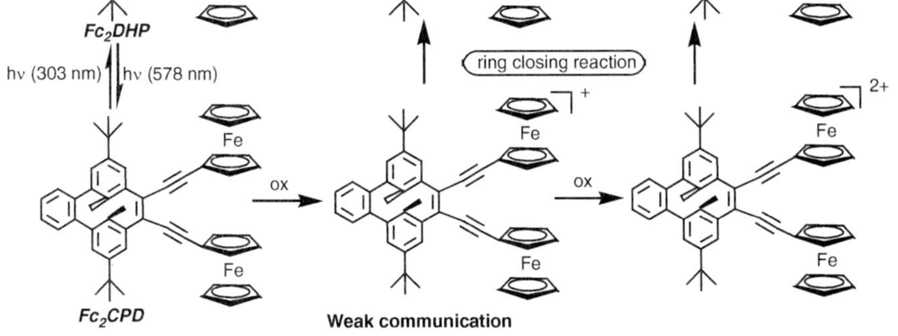

7 Other Photochromic Metal Complexes

A photochromic dianthryl molecule can act as a triplet energy transfer quencher when combining with a ruthenium diimine complex [123]. UV excitation of the systems (X = CH$_2$O, C(O)O) results in a photoinduced cycloaddition of the dianthryl unit, leading to an increase of the luminescence intensity. Near-field optical addressing of this luminescent photoswitcheable system (X = CH$_2$O) as a dopant in PMMA films has been demonstrated [124] (Fig. 14).

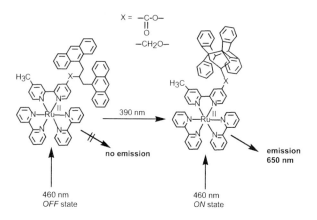

This work is based on work previously reported by Belser and De Cola on a Re complex in which the bipyridine ligand incorporates an anthryl derivative [125].

Fig. 14 *Black lines*: absorption and emission (λ_{exc} = 360 nm) spectra measured for Ru-anthryl (X = CH$_2$O) for dispersed in a 45 ± 5 nm thick PMMA film. Absorption peaks associated with the dianthryl unit are indicated by *asterisk*. *Red lines*: analogous spectra measured following the formation of the cycloadduct by 400 nm irradiation (10 mW, 5 min) of a ~18 mm diameter circular area of the doped PMMA film. (Reprinted with permission from [124])

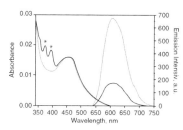

7.1 Terthiazole Derivatives

The luminescence of the Eu(III) complex Eu(THIA)$_2$(HFA)$_3$ (THIA: 4,5-bis (5-methyl-2-phenylthiazole) is efficiently quenched when the photochromic ligand THIA is converted from the colorless ring-open form to the closed-ring form upon UV light irradiation, resulting in change of the emission intensity [126].

8 Conclusion

A number of systems where photoresponsive molecules are used as ligands to form metal complexes, mono- or multimetallic, have been reported during the last decade. Such complexes exist for a wide range of metal centers (Cr, Mo, Mn, Re, Fe, Ru, Os, Co, Rh, Ir, Ni, Pt, Cu, Ag, Au, Zn) and more recently lanthanides (Eu). The photochromic properties of the organic framework were found, in some cases, to be enhanced by complexation and, as its organic counterpart, also effective in single-crystalline phase.

These reports shows that the photochromic properties could be readily tuned by changing the nature of the metal system without modifying the photochromic ligand. Thus, by a rational design of metal centers, ancillary ligands, and counter-anions, photochromic behavior can be controlled or modified. In addition, triplet reaction pathways have been demonstrated, allowing the extension of the excitation wavelengths to lower energies, which are less destructive. Finally, these metal systems allow the photomodulation of various properties such as luminescence, second-order nonlinear optics, and magnetic or electronic interaction for which the presence of the organometallic fragment is crucial for the properties mentioned. These investigations were initiated 10 years ago and yet this field still has to be explored.

References

1. Feringa BL (ed) (2001) Molecular switches. Wiley, Weinheim, Germany
2. Dürr H, Bouas-Laurent H (1990) Photochromism: molecules and systems. Elsevier, Amsterdam
3. Irie M (2000) Special issue on photochromism: memories and switches. Chem Rev 100:1683–1890 and references therein
4. Raymo FM, Tomasulo M (2005) Chem Soc Rev 34:327–336
5. Raymo FM, Tomasulo M (2005) J Phys Chem A 109:7343–7352
6. Kumar GS, Neckers DC (1989) Chem Rev 89:1915–1925
7. Natansohn A, Rochon P (2002) Chem Rev 102:4139–4175
8. Delaire JA, Nakatani K (2000) Chem Rev 100:1817–1845
9. Kawata S, Kawata Y (2000) Chem Rev 100:1777–1788
10. Kurihara M, Nishihara H (2002) Coord Chem Rev 226:125–135
11. Nishihara H (2004) Bull Chem Soc Jpn 77:407–428
12. Kume S, Nishihara H (2008) Dalton Trans 3260–3271
13. Nesmeyanov AN, Perevalova EG, Nikitina TV (1961) Dokl Akad Nauk SSSR 138:118
14. Nesmeyanov AN, Sazonova VA, Romanenko VI (1961) Dokl Akad Nauk SSSR 157:992
15. Kurihara M, Matsuda T, Hirooka A, Yutaka T, Nishihara H (2000) J Am Chem Soc 122:12373–12374
16. Kurihara M, Hirooka A, Kume S, Sugimoto M, Nishihara H (2002) J Am Chem Soc 124:8800–8801
17. Sakamoto A, Hirooka A, Namiki K, Kurihara M, Murata M, Sugimoto M, Nishihara H (2005) Inorg Chem 44:7547–7558
18. Namiki K, Sakamoto A, Murata M, Kume S, Nishihara H (2007) Chem Commun 4650–4652
19. Muraoka T, Kinbara K, Kobayashi Y, Aida T (2003) J Am Chem Soc 125:5612–5613
20. Muraoka T, Kinbara K, Aida T (2007) Chem Commun 1441–1443
21. Muraoka T, Kinbara K, Aida T (2006) Nature 440:512–515
22. Kinbara K, Muraoka T, Aida T (2008) Org Biomol Chem 6:1871–1876
23. Kume S, Kurihara M, Nishihara H (2001) Chem Commun 1656–1657
24. Yamaguchi K, Kume S, Namiki K, Murata M, Tamai N, Nishihara H (2005) Inorg Chem 44:9056–9067
25. Viau L, Bidault S, Maury O, Brasselet S, Ledoux I, Zyss J, Ishow E, Nakatani K, Le Bozec H (2004) J Am Chem Soc 126:8386–8387
26. Maury O, Le Bozec H (2004) Acc Chem Res 38:691–704
27. Viau L, Malkowsky I, Costuas K, Boulin S, Toupet L, Ishow E, Nakatani K, Maury O, Le Bozec H (2006) ChemPhysChem 7:644–657
28. Bidault S, Viau L, Maury O, Brasselet S, Zyss J, Ishow E, Nakatani K, Le Bozec H (2006) Adv Funct Mater 16:2252–2262
29. Kume K, Kurihara M, Nishihara H (2003) Inorg Chem 42:2194–2196
30. Kume K, Murata M, Ozeki T, Nishihara H (2004) J Am Chem Soc 127:490–491
31. Yutaka T, Kurihara M, Nishihara H (2000) Mol Cryst Liq Cryst 343:193–198
32. Yutaka T, Kurihara M, Kubo K, Nishihara H (2000) Inorg Chem 39:3438–3439
33. Yutaka T, Mori I, Kurihara M, Tani M, Kubo K, Furusho S, Matsumura K, Tamai N, Nishihara H (2001) Inorg Chem 40:4986–4995
34. Yutaka T, Mori I, Kurihara M, Tamai N, Nishihara H (2003) Inorg Chem 42:6306–6313
35. Yutaka T, Mori I, Kurihara M, Mizutani J, Tamai N, Kawai T, Irie M, Nishihara H (2002) Inorg Chem 41:7143–7150
36. Nihei M, Kurihara M, Mizutani J, Nishihara H (2001) Chem Lett 852–853
37. Nihei M, Kurihara M, Mizutani J, Nishihara H (2003) J Am Chem Soc 125:2964–2973
38. Sakamoto R, Murata M, Kume S, Sampei H, Sugimoto M, Nishihara H (2005) Chem Commun 1215–1217

39. Cummins SD, Eisenberg R (1995) Inorg Chem 34:2007–2014
40. Lang H, George DSA, Rheinwald G (2000) Coord Chem Rev 206/207:101–197
41. Tang HS, Zhu N, Yam VWW (2007) Organometallics 26:22–25
42. Luc J, Bouchouit K, Czaplicki R, Fillaut J-L, Sahraoui B (2008) Opt Express 16: 15633–15639
43. Luc J, Niziol J, Sniechowski Sahraoui B, Fillaut J-L, Krupka O (2008) Mol Cryst Liq Cryst 485:242–253
44. Irie M (2000) Chem Rev 100:1685–1716
45. Tian H, Yang S (2004) Chem Soc Rev 33:85–97
46. Zhou Z, Yang H, Shi M, Xiao S, Li F, Yi T, Huang C (2007) ChemPhysChem 8: 1289–1292
47. Zhou Z, Xiao S, Xu J, Zhiqiang L, Shi M, Li F, Yi T, Huang C (2006) Org Lett 8: 3911–3914
48. Zhou Z, Hu H, Yang H, Yi T, Huang K, Yu M, Li F, Huang C (2008) Chem Commun 4786–4788
49. Golovka TA, Kozlov DV, Neckers DC (2005) J Org Chem 70:5545–5549
50. Matsuda K, Takayama K, Irie M (2001) Chem Commun 363–364
51. Matsuda K, Takayama K, Irie M (2004) Inorg Chem 43:482–489
52. Gilat SL, Kawai SH, Lehn JM (1995) Chem Eur J 1:275–284
53. Qin B, Yao R, Zhao X, Tian H (2003) Org Biomol Chem 1:2187–2191
54. Samachetty HD, Branda NR (2005) Chem Commun 2840
55. Samachetty HD, Branda NR (2006) Pure Appl Chem 78:2351–2359
56. Fernandez-Acebes A, Lehn JM (1998) Adv Mater 10:1519–1522
57. Fernandez-Acebes A, Lehn JM (1999) Chem Eur J 5:3285–3291
58. Senechal-David K, Zaman N, Walko M, Halza E, Riviere E, Guillot R, Feringa BL, Boillot ML (2008) Dalton Trans 1932–1936
59. Matsuda K, Shinkai Y, Irie M (2004) Inorg Chem 43:3774–3776
60. Qin B, Yao R, Tian H (2004) Inorg Chim Acta 357:3382–3384
61. Munakata M, Han J, Nabei A, Kuroda-Sowa T, Maekawa M, Suenaga Y, Gunjima N (2006) Inorg Chim Acta 359:4281–4288
62. Han J, Konada H, Kuroda-Sowa T, Maekawa M, Suenaga Y, Isihara H, Munakata M (2006) Inorg Chim Acta 359:99–108
63. Munakata M, Wu LP, Kuroda-Sowa T, Maekawa M, Suenaga Y, Furuichi K (1996) J Am Chem Soc 118:3305–3306
64. Konada H, Wu LP, Munakata M, Kuroda-Sowa T, Maekawa M, Suenaga Y (2003) Inorg Chem 42:1928–1934
65. Myles AJ, Branda NR (2003) Macromolecules 36:298–303
66. Han J, Maekawa M, Suenaga Y, Ebisu H, Nabei A, Kuroda-Sowa T, Munakata M (2007) Inorg Chem 46:3313–3321
67. Giraud M, Léaustic A, Guillot R, Yu P, Lacroix PG, Nakatani K, Pansu R, Maurel F (2007) J Mater Chem 17:4414–4425
68. Jukes RTF, Adamo V, Hartl F, Belser P, De Cola L (2004) Inorg Chem 23:2779–2792
69. Jukes RTF, Adamo V, Hartl F, Belser P, De Cola L (2005) Coord Chem Rev 249: 1327–1335
70. Belser P, De Cola L, Hartl F, Adamo V, Bozic B, Chriqui Y, Iyer VM, Jukes RTF, Kühni J, Querol M, Roma S, Sallucc N (2006) Adv Funct Mater 16:195–208
71. Zhong YW, Vila N, Henderson JC, Flores-Torres S, Abruna HD (2007) Inorg Chem 46:104470–10472
72. Yam VWW, Ko CC, Zhu N (2004) J Am Chem Soc 126:12734–12735
73. Ko CC, Kwok WM, Yam VWW, Phillips DL (2006) Chem Eur J 12:5840–5848
74. Lee JKW, Ko CC, Wong KMC, Zhu N, Yam VWW (2007) Organometallics 26:12–15
75. Belser P, Adamo V, Kühni J (2006) Synthesis 12:1946–1948
76. Xiao S, Yi T, Zhou Y, Zhao Q, Li F, Huang C (2006) Tetrahedron 62:10072–10078

77. Indelli MT, Carli S, Ghirotti M, Chiorboli C, Ravaglia M, Garavelli M, Scandola F (2008) J Am Chem Soc 130:7286–7299
78. Lee PHM, Ko CC, Zhu N, Yam VWW (2007) J Am Chem Soc 129:6058–6059
79. Aubert V, Guerchais V, Ishow E, Hoang-Thi K, Ledoux I, Nakatani K, Le Bozec H (2008) Angew Chem Int Ed 47:577–580
80. Takayama K, Matsuda K, Irie M (2003) Chem Eur J 9:5605–5609
81. Tanake Y, Inagaki A, Akita M (2007) Chem Commun 1169
82. Liu Y, Lagrost C, Costuas C, Tchouar N, Le Bozec H, Rigaut S (2008) Chem Commun 6117–6119
83. Motoyama K, Koike T, Akita M (2008) Chem Commun 5812–5814
84. Fraysse S, Coudret C, Launay JP (2000) Eur J Inorg Chem 1581–1590
85. Guirado G, Coudret C, Launay JP (2007) J Phys Chem C 111:2770–2776
86. Carella A, Coudret C, Guirado G, Rapenne G, Vives G, Launay JP (2007) Dalton Trans 177–186
87. Uchida K, Inaga A, Akita M (2007) Organometallics 26:5030–5041
88. Sud D, McDonald R, Branda NR (2005) Inorg Chem 44:5960–5962
89. Murguly E, Norsten TB, Branda NR (2001) Angew Chem Int Ed 40:1752–1755
90. Kawai T, Sasaki T, Irie M (2001) Chem Commun 711–712
91. Tian H, Chen B, Tu H, Müllen K (2002) Adv Mater 14:918–923
92. Ko CC, Lam WH, Yam VWW (2008) Chem Commun 5203–5205
93. Nakagawa T, Atsumi K, Nakashima T, Hasegawa Y, Kawai T (2007) Chem Lett 36:372–373
94. Kim HJ, Jang JH, Choi H, Lee T, Ko J, Yoon M, Kim HJ (2008) Inorg Chem 47:2411–2415
95. Jung I, Choi H, Kim E, Lee C-H, Kang S, Ko J (2005) Tetrahedron 12256–12263
96. Crano JC, Guglielmetti RJ (1999) Organic photochromic and thermochromic compounds, vol 1. Kluwer, Dordecht
97. Minkin VI (2004) Chem Rev 104:2751–2776 and references therein
98. Fischer E, Hirshberg Y (1952) J Chem Soc 4522–4524
99. Nagashima S, Murata M, Nishihara H (2006) Angew Chem Int Ed 45:4298–4301
100. Bahr JL, Kodis G, de la Garza L, Lin S, Moore AL, Moore TA, Gust D (2001) J Am Chem Soc 123:7124–7133
101. Yam VWW, Ko CC, Wu LX, Wong KMC, Cheung KK (2000) Organometallics 19:1820–1822
102. Ko CC, Wu LX, Wong KMC, Zhu N, Yam VWW (2004) Chem Eur J 10:766–776
103. Bao Z, Ng K-Y, Yam VWW, Ko CC, Zhu N, Wu L (2008) Inorg Chem 47:8912–8920
104. Querol M, Bozic B, Salluce N, Belser P (2003) Polyhedron 22:655–664
105. Jukes RTF, Bozic B, Hartl F, Belser P, De Cola L (2006) Inorg Chem 45:8326–8341
106. Khairutdinov RF, Giertz K, Hurst JK, Voloshina EN, Voloshin NA, Minkin VI (1998) J Am Chem Soc 120:12707–12713
107. Kopelman RA, Snyder SM, Frank NL (2003) J Am Chem Soc 125:13684–13685
108. Moriuchi A, Uchida K, Inagaki AA, Akita M (2005) Organometallics 24:6382–6392
109. Miyashita A, Iwamoto A, Kuwayama T, Aoki Y, Hirano M, Nohira H (1997) Chem Lett 965–966
110. Wojtyk JTC, Kazmaier PM, Buncel E (1998) Chem Commun 1703–1704
111. Wojtyk JTC, Kazmaier PM, Buncel E (2001) Chem Mater 13:2547–2551
112. Chisibov AK, Görner H (1998) Chem Phys 237:425–442
113. Anguille S, Brun P, Guglielmetti R (1998) Heterocycl Commun 4:63
114. Anguille S, Brun P, Guglielmetti R, Strokach YP, Ignatin AA, Barachevsky VA, Alfimov MV (2001) J Chem Soc Perkin Trans 2:639–644
115. Brun P, Guglielmetti R, Anguille S (2002) 16:271–276
116. Mitchell RH (1999) Eur J Org Chem 2695–2703

117. Mitchell RH, Ward TR, Chen Y, Wang Y, Weerawarna SA, Dibble PW, Marsella MJ, Almutairi A, Wang Z-Q (2003) J Am Chem Soc 125:2974–2988
118. Mitchell RH, Brkic Z, Berg DJ, Barclay TM (2002) J Am Chem Soc 124:11983–11988
119. Mitchell RH, Brkic Z, Sauro VA, Berg DJ (2003) J Am Chem Soc 125:7581–7585
120. Fan W, Berg DJ, Mitchell RH, Barclay TM (2007) Organometallics 26:4562–4567
121. Mitchell RH, Fan W, Lau DYK, Berg DJ (2004) J Org Chem 69:549–554
122. Muratsugu S, Kume S, Nishihara H (2008) J Am Chem Soc 130:7204–7205
123. Tyson DS, Bignozzi CA, Castellano FN (2002) J Am Chem Soc 124:4562–4563
124. Ferri V, Scoponi M, Bignozzi CA, Tyson DS, Castellano FN, Redmond G (2004) Nanoletters 4:835–839
125. Beyeler A, Belser P, De Cola L (1997) Angew Chem Int Ed Engl 36:2779–2781
126. Nakagawa T, Atmusi K, Nakashima T, Hasegawa Y, Kawai T (2007) Chem Lett 36:372–373

Index

A
Absorption, nonlinear 57
Acetylide/vinylidene pairs 27
Alkynyl ligands 1, 26
Alkynyl ruthenium dendrimers 68
Amines, ligands 1
Amino acids 161
Anion sensors 152
Aryldiazovinylidene 27
2-Arylpyridines 93
Au(I) 151
Avidin 163
Azobenzene 174
Azobenzene-conjugated bipyridine ligands 178
– terpyridine ligands 182
Azobenzodithiolenes 184
Azo-containing metal complexes, photochromic 173
Azoferrocene 173

B
Benzodithiolethione 184
Benzopyran 174
Bimetallic complexes 42
Biological labeling 131
– iridium complexes 113
Biomolecules, large, sensing 161
Bipyridines 1
– azobenzene-conjugated 178
Biquinoline 87
Bis(pyrazolyl)borates 94
Bis(salicylaldiminato) framework 21
Blue emitters 128
BODIPY 190
Bovine serum albumin (BSA) 163

C
Carbazole, norbornene-appended 98
Carboxylate ligands, DTE-based 191
Cation sensors 145
– ferrocene 149
Chelating ligands 1, 13
Chemosensors, iridium complexes 113
Cisplatin 75
Colorimetry 143
Coordination complexes 1, 7
Cubic nonlinearity 57
Cyanide ligands 147
Cyano ligands, DTE-based 191
Cyano-Os(II) complexes 147
Cyclometallated ligands 1, 29, 93
– N^C= bidentate 93
Cyclopentadienyl (Cp) ligands 25, 148
Cyclophanediene (CPD) 218

D
Degenerate four-wave mixing (DFWM) 65
Dendrimeric species 18, 115
Diamidino-2-phenylindole dihydrochloride (DAPI) 133
Diarylethene (DTE) derivatives 188
4,4-Difluoro-4-bora-3a,4a-diaza-s-indacene 190
Diimine Os(II) 143

Diimine Ru(II) 143
Dimethyldihydropyrene 173, 218
1,2-Dithienylcyclopentene 188
Dithienylcyclopentene, organo-boron DTE-based 189
Dithienylethene (DTE) 171, 174, 188
DNA, detection 161
DTE derivatives, metal-based, photo-/electrochromic properties 205
DTE-based ligands, catalysis 208

E
Electric field induced second harmonic generation (EFISH) 5
Electroluminescence 77
– platinum compounds 75
Erbium 133
Estrodiol 164
Europium 133
Excimers 96
Excited states, tuning 81
Excitons 78

F
Ferrocene 143
Ferrocene–azobenzene 173, 176
Ferrocenylspiropyran 211
Fluorescence 26, 143
Four-wave mixing 60
Fructose, detection 161

G
Glucose, optical detection, Re(I) complexes 161
Glutathione 161
Gold(I) complexes 151

H
Homocysteine 161
Hyper-Rayleigh scattering (HRS) 5

I
Indium–tin–oxide (ITO) 126, 176
Intraligand charge transfer (ILCT) 8, 146

Iridium complexes 154
– biological labeling 113
– cyclometallated 30, 143, 147
– ionic 118, 123
– luminescent, applications 126
– neutral 120

J
Jabłoński diagram 77

L
Lanthanide complexes 20
Lanthanide luminescence, sensitizer 133
Ligands, photochromic 171
Light-driven chiral molecular scissors 176
Luminescence, iridium complexes 113
– photoregulation 198
– platinum compounds 75, 79

M
Macrocyclic ligands 1, 31
Magnetic interaction, photoswitching 204
Memories 171
Metal alkynyl complexes, azobenzene-containing 186
Metal complexes, 1,2-dithienylethene 188
Metallacycles 86
Metalladithiolenes, azobenzene groups 184
Metallocenes 1, 23
Metallocenyl benzopyran derivatives, photochromic 217
Metal-to-ligand charge transfer (MLCT) 143
Methionine 161
Molecular recognition 143
Molecular scissors, light-driven chiral 176
Molecular sensors 143
Molecular switches 57
Multi-DTE metal complexes 209
Multifunctional complexes 97

N
Neodymium 133
Neutral Iridium complex 114
NLO, second-order 1
– third-order 58

Index

Nonlinear absorption 57
Nonlinear refraction 57
Nonlinearity, cubic 57

O
OLEDs 75, 113
– applications 126
– electroluminescence 77
– iridium complexes 113
– platinum compounds 75
Open/closed motion 177
Optically-transparent thin-layer electrochemical (OTTLE) cell 70
Organic light-emitting diodes (OLEDs) 113
Organic molecular materials 6
Organic molecules, small, in solution 159
Organic photochromic molecules 171
Organometallic carbon, X= 85
Organometallic complexes 1, 7
Organometallic sensors, fluorogenic/chromogenic 143
Organometallics, third-order NLO 58
Osmium(II) complexes 143, 147
Osmocenyl 217

P
Phenanthrolines 1, 15, 87
2-Phenylpyridine, cyclometallated 31
Phosphorescence 26
Photochemistry, platinum compounds 75
Photochromes 171
Photochromic azo-containing metal complexes 173
Photochromic metal complexes 220
Photophysical properties 120
Photoswitching, second-order NLO activity 203
Phthalocyanines 1, 39
Platinum, luminescence 75
Platinum(II) complexes 150, 154
– colour tuning 93
– N^C^N-binding ligands 104
– N^N^C-binding ligands 100
– N^N^O-binding ligands 104
Platinum(II) porphyrins 81
Pliers, light-powered molecular 176

Polypyridine metal complexes, spiropyran-/spirooxazine 212
Porphyrin spiropyran metal complexes 212
Porphyrins 1
Proteins, detection 163
Pseudo-cyclometallates 89
Pseudotetrahedral Cu(I) 179
Pt-alkynyls 143
Pt(II) alkynyl terpyridine 150
Pt(quinolinol)$_2$ 96
PtL$_2$X$_2$ complexes 82
Pyrazolate derivatives 116
2-Pyrenylacetylide ligand 101
Pyridine ligands, DTE-based 191
4-Pyridinecarboxaldehydeazine 147
Pyridines, ligands 1
Pyridyl polyene chromophores 9
Pyrroles 91

R
Re(I) tricarbonyl 152
Re(I)–Pd(II) square 152
Red emitters 128
Refraction, nonlinear 57
Reporters 143
Rhenium(I) carbonyl complexes 145, 152
Ruthenium(II) complexes 147, 154
Ruthenocenyl 217

S
Salen, L$_2$X$_2$=87
Salicylaldehydes 21
Schiff bases 1
Scissors, light-driven chiral molecular 176
Second-order NLO 1, 3
– activity, photoswitching 203
Sensor applications 135
Sensors, anion 152
– cation 145
– ferrocene-based 148, 155
Small molecule detection 156
Spectral dependencies 67
Spirobenzopyrans 216
– [(Z6-Spirobenzopyran)Ru(C$_5$Me$_5$)], diastereomeric isomerism 216
Spironaphthoxazine 214

Spirooxaxine (SO) 174
 – metal complexes, photochromic 210
Spiropyran (SP) 174
 – metal ions 216
Spiropyran-containing metal complexes, photochromic 210
Stilbazoles, ligands 1
Stilbene (trans)-1-ferrocenyl-2-(4-nitrophenyl)ethylene 23
Sugar, detection 161
Sulfhydryl amino acids 161
Switches 69, 171
 – molecular 57

T
Terdentate ligands 99
Terpyridine ligands, azobenzene-conjugated 182
Terpyridines 1
Terthiazole derivatives 221
Thiazinane group 161
Third-order NLO 58
4-Tolylazophenyl-2,2'-bipyridine 178
Transition metal complexes 171

1,3,5-Triamino-2,4,6-trinitrobenzene 7
Tricarbonyl Re(I) complexes 145
Tris(2,2'-bipyridyl)ruthenium(II) cation 144
Tris(styrylbipyridine)zinc(II) 178
TRISPHAT 16

U
Up-conversion rare-earth nanophosphors (UCNPs) 190
UV-vis spectroscopy 143

V
Vinylidene ligands 1, 26
Volatile organic compounds, detection 156

W
WOLEDs (white light-emitting devices) 96, 129

Y
Ytterbium 133